AF772109

Dear Jim.

Thank you so much for being my PR mentor in Austin.

I really appreciate our long-term friendship.

Kjell Brataas

Crisis Communication

Crisis Communication is an in-depth examination of recent tragedies and natural disasters that have occurred around the globe.

The book covers three types of incidents: natural catastrophes, accidents and terror attacks. It focuses on the communication aspect of each incident and provides accounts from people handling the event. Each chapter offers a detailed description of the event and supplementary facts and illustrations from a variety of sources.

With a focus on critical communication elements and lessons learned, Brataas offers valuable advice – based on personal experience with natural disasters, accidents and terror attacks – on some of the most effective ways to prepare for and deal with a crisis. Topics range from interview situations and social media to victim support and active shooter events.

This book will be invaluable to those working in public relations and communications, as well as to those working with human resources and general management.

Kjell Brataas has been on the front lines of crisis communication during some of Norway's most devastating crises. He held central positions after the tsunami in 2004 and following the terrorist attacks in Oslo and on Utoya on July 22, 2011. Brataas has practical experience from private companies and government ministries. He has presented at crisis communication conferences around the world, including London, Toronto, Istanbul and Denver, and he has lectured on crisis communication at the university level.

Crisis Communication

Case Studies and Lessons Learned from International Disasters

Kjell Brataas

Routledge
Taylor & Francis Group
NEW YORK AND LONDON

First published 2018
by Routledge
711 Third Avenue, New York, NY 10017

and by Routledge
2 Park Square, Milton Park, Abingdon, Oxon, OX14 4RN

Routledge is an imprint of the Taylor & Francis Group, an informa business

Library of Congress Cataloging-in-Publication Data
Names: Brataas, Kjell, author.
Title: Crisis communication : case studies and lessons learned
from international disasters / Kjell Brataas.
Description: 1 Edition. | New York : Routledge, 2018. |
Includes bibliographical references and index.
Identifiers: LCCN 2017037478 (print) | LCCN 2017057230 (ebook) |
ISBN 9781315368245 (Master) | ISBN 9781498751353 (WebPDF) |
ISBN 9781315351285 (ePub) | ISBN 9781315332246 (Mobipocket/Kindle) |
ISBN 9781498751346 (hardback : alk. paper) | ISBN 9781315368245 (ebk)
Subjects: LCSH: Crisis management. | Emergency management. |
Communication in management. Classification: LCC HD49 (ebook) |
LCC HD49.B73 2018 (print) | DDC 658.4/77–dc23
LC record available at https://lccn.loc.gov/2017037478

ISBN: 978-1-498-75134-6 (hbk)
ISBN: 978-1-315-36824-5 (ebk)

Typeset in Sabon
by Out of House Publishing

MIX
Paper from
responsible sources
FSC™ C013985
FSC
www.fsc.org

Printed in the United Kingdom
by Henry Ling Limited

Contents

Figures

About the Author

For most of his career, Kjell Brataas has been employed by different ministries within the Norwegian government. He soon displayed an interest in crisis communication and volunteered his skills to a support group within the government that was to be summoned in the case of a major crisis. Brataas had a central role in the Norwegian government's follow-up of the tsunami in 2004, including weekly meetings with next-of-kin and arranging for two trips to Thailand for family members in 2005. When an official report on the tragedy suggested a formal group be established in the Ministry of Justice, Brataas was asked to be in charge of setting up and designing crisis communication capability within the group.

In July 2011, Brataas was called in to handle international media, internal communication and next-of-kin support after the bombing in Oslo and the mass shooting on the island of Utoya. He assisted at the media center and monitored social media during visits to Utoya in August for survivors and family members.

Brataas studied journalism and PR at the University of Texas at Austin. He has lectured and shared his knowledge and experience in a variety of settings around the world. He has taught multiple courses in crisis communication at the Emergency Planning College of the Ministry of Justice, and he is a guest lecturer at the Arctic University of Norway. Additionally, Brataas has presented at the International Air Transport Association (IATA) conference on crisis communication and social media in Istanbul, the Intermedix Summit in Denver and the World Conference on Disaster Management (WCDM) in Toronto (2014–2016). In addition to serving as a presenter, Brataas has been a member of the

Advisory Board for WCDM, which included determining speakers and organizing attendee events.

In addition to his work, Brataas enjoys traveling, skiing and hiking in the mountains. He lives in Billingstad outside of Oslo with his wife Janelle and their three children.

The author welcomes comments and suggestions. You can contact Kjell Brataas through email: kjell@brataascrisis.com.

Foreword

It is an absolute pleasure to be asked to write a foreword to this excellent book by Kjell Brataas, whom I first met several years ago in Canada during the annual World Conference on Disaster Management. It soon became apparent that Kjell is a gifted and highly competent communicator himself, and it is no surprise that he quickly became a key member of the WCDM Advisory Board. His personal experience with a variety of crisis communication issues, media handling, hotlines for victims, support groups, site visits, memorial services and so on have served him well as the author of this important book, which focuses on disaster communication with vital case studies and lessons learned from international disasters to avoid repeating past mistakes. This point was first made way back in 1906 by the Spanish philosopher George Santayana, who presciently stated then, as nowadays, that 'Those who cannot remember the past are condemned to repeat it.'

If I think about my own experience in disaster management, stretching back over nearly four decades, I am reminded of one particular headline (which presently attracts over 29 million hits on Google): 'Crisis. What crisis?' So said British Prime Minister Jim Callaghan way back in 1979, returning from sunny Guadeloupe to a damp, disastrous and strike-ridden U.K. with rubbish piled high on street corners and even the dead unburied in some cities. Or so history says. The fact is, Callaghan never actually said these words. Instead, a reporter working for the popular *Sun* newspaper used them as a headline that caught the popular impression of an out-of-touch government caught in the headlights. Perhaps 'PM plays down problems' might have been a more accurate headline, but 'Crisis? What crisis?' suited the mood of the nation and has since become part of British political folklore. The impact of this miscommunication was so forceful that a few months later Callaghan's government was voted out of office.

Among the many topics that underpin effective disaster management, I cannot find a more critical heading to focus on than communication. Good leaders and supportive crisis teams are all very well, but if they cannot communicate properly or are deliberately misquoted, their efforts become pointless. But communication is not just about command: I am delighted to note that the largest chapter in this book is about family and victim support, dealing with an often forgotten or neglected aspect of crisis management.

We now live in a world where the extraordinary has become commonplace and the unexpected is now regularly anticipated. Add to that hundreds of predatory news organizations, immediate and global communications, stories of abandoned disaster victims and hitherto steadfast organizations frequently discredited and ridiculed, and you might be correct to assume that we are perhaps more vulnerable to all aspects of communication in a disaster than ever before. This also means, of course, that we are more aware of crises, yet at the same time more unforgiving if those trying to resolve crises do not deliver the solutions we have been led to expect: an extremely difficult, if not impossible, challenge for any government, organization or body of people tasked with managing any crisis, in that, for example, the speed of social communication would have been unimaginable even ten years ago.

Nowadays we exist in an increasingly fragile, bewildering and interconnected society, where just about all essential services we rely on are far more entangled than we realize. When something goes wrong, the consequences are therefore more sudden and widespread, sometimes made worse by secrecy, scapegoats and silos: we cannot be told, we need someone to blame and in any case, we work separately.

If we fail to share assumptions and ideas on disaster communication between organizations, sectors, regions and even countries, we must surely prepare to fail in the future. It's therefore time to learn from the lessons outlined in this book, climb much further out of our silos and dismantle some of the unnecessary boundaries that exist, especially in an age of unparalleled and instant communication that, not forgetting simple acts of human kindness, we should use to our advantage whenever the next disaster strikes.

This up-to-date and highly readable book, which deliberately focuses on international crises, could not be better timed. Never has it been more important to dismantle borders, learn from each other, take better care of victims and above all, learn to be better communicators.

Peter Power BA FBCI FIRM JP
Chairman WCDM
Co-author UK Government standard on crisis management (BS 11200)
Managing Director, Visor Consultants

Preface

On December 14, 1911, the Norwegian explorer Roald Amundsen and his team accomplished an astonishing achievement – they were the first human beings to reach the South Pole. This feat would fascinate a world hungry with questions; how was it humanly possible, why did they succeed and not fail, and what hardships did they endure during their journey? All of those questions would be answered, but it would be almost three months before the first news article appeared to inform the world about Amundsen's success.

In so many respects, the world today is totally different than it was in Amundsen's time. The *speed* at which news and information are conveyed, analyzed and disseminated has shrunk from taking months to being instantaneous. Smartphones with cameras can broadcast live from almost anywhere on earth, and how governments and businesses deal with a crisis and respond appropriately is judged immediately on television and social media. The days of spending hours agonizing over a press statement are gone. Today's world must factor in citizen journalists, live feeds, hashtags and crowdsourced information – and these transformations are here to stay.

As the possibilities in communication evolve, crisis management has also had to adapt: Not just because there is an expectation of instant communication, but because of greater demands for immediate answers, more pressure for transparency and an expectation of openness. These changes are mostly positive, but it is essential that government officials, business leaders and communications professionals alike realize and accept this new reality. They must be prepared for it and embrace it. That is the purpose of this book.

MY STORY

I have always enjoyed writing. Growing up in Norway, I earned a bit of extra income writing articles and conducting interviews for magazines and newspapers, and as a young adult I published a few travel guides. However, writing a book in a foreign language was not one of my aspirations.

My first personal encounter with managing a disaster came in 2004, when the tsunami hit Asia and killed close to 230,000 people, including 84 Norwegians. At the time, I was a senior communications advisor with the Norwegian Ministry of Health, and I was tasked with a variety of crisis communication responsibilities. People from more than 50 countries died in the tragedy, and I had the opportunity to see how colleagues from other countries communicated about the crisis and to pick up some best practices and lessons learned.

That event did not involve interacting with social media. Facebook was still in its infancy, and YouTube, Twitter, Snapchat and a host of other social media platforms had not yet been invented. But social media played a significant role in the next crisis management effort I was involved with. On July 22, 2011, news about a bombing in downtown Oslo that killed eight people quickly spread through Facebook and Twitter as well as traditional news outlets. Almost at the same time as that event was unfolding, young people at a youth camp on Utoya Island – more than 23 miles from Oslo – were sending terrified messages *in real time* about a mass shooting taking place on the island.

Another interest of mine is meeting new people. My career and crisis experience have taken me to many countries and continents where I have had the opportunity to speak at professional seminars, to share ideas and to learn from friends and colleagues. Attending these conferences has taught me that there is a large group of professionals out there who have varied experience and knowledge, and from whom there is much to learn. Sadly, though, working in the aftermath of disasters has also forced me to meet and confront the human side of tragedy, teaching me that the greatest toll is on the victims, their families and those who knew and loved them. It is our job, as those tasked with the management of a crisis, to do so with as much compassion, professionalism and speed as possible.

THIS BOOK

This book includes several examples of how crisis communication can make a tragic situation even worse. A common mistake is when a government official or the chief executive officer of a company chooses words badly or does not focus on the victims. Many people remember the CEO of British Petroleum, in response to a reporter about the disaster that killed 11 people on the offshore drilling platform *Deepwater Horizon*, saying, 'I want my life back.' His tone-deaf response infuriated the victims' families and the general public alike. Another example of atrocious communication is recalled from the maritime disaster in 1994, when MS *Estonia* sank in the middle of the night in the Baltic Sea and 852 souls perished. The first press release put out by the ship's owners created a firestorm of fury when they declared, 'We can confirm that the ship was insured.'

There are other lessons to be learned as well. One has to do with the care taken when handling *lists* and *numbers*. It is essential to keep track of who has died, where victims are, which hospitals have admitted injured, where next-of-kin should report, and contact information for both the next-of-kin and the media. Accurate dissemination of information, and doing this in a timely fashion, is among the most demanding tasks of crisis management. It can mean the difference between staying on top of a crisis and letting events get ahead of you and out of control.

Several chapters in this book cover the technical and theoretical aspects of crisis communication – these are the 'nuts and bolts' of our work. And I include sections on working with the media and other communication channels and recommendations for handling databases. But it is important to emphasize that disaster management and crisis communication are, first and foremost, about relating to people. As the result of a disaster, countless victims have their lives changed forever, and it is paramount to keep that in mind when you draft press releases, comment on social media or give interviews to print and electronic media outlets. A poor response or ill-considered thought blurted out on CNN or BBC can be devastating to your organization – but even more so to the victims and their families.

For communicators as well as managers, it is also important to understand the psychological aspects of being in a crisis situation, or what it is like to be the sudden victim of a tragic event. Initial shock often turns to anger, but effective and direct intervention by management, combined with professional communication, can go a long way towards mitigating the situation.

FOCUS ON COMMUNICATION

My main emphasis in this book is on the communication aspects of handling a disaster, including the way we speak, write, tweet or post about events during and after an event. The book is divided into two parts. The first is called 'Case Studies' and includes narratives of natural disasters, terror attacks and disasters in the transportation sector. I have included 11 different case studies, and I have chosen them because they – communications-wise – were handled exceptionally well or exceptionally poorly, or because they involved new and modern ways of communicating. Two of the case studies include a personal voice, as I was heavily involved in the handling of the tsunami in 2004 and the terror in Oslo and Utoya in 2011. The second part of the book focuses on 'Lessons Learned' and best practices, many of which are drawn from case studies presented earlier. Most of the advice given is based on my personal experience and knowledge, but I also include hints and advice from additional sources and refer to studies and research from people and organizations all over the world.

In addition, three sub-headings should be mentioned, as they are featured widely throughout the book:

Social media have profoundly changed the way crisis management works and how we, as crisis managers, do our jobs. I focus on platforms such as Facebook, Twitter, YouTube and others. In fact, because of their importance, some of the case studies deal mostly with the use of social media. While it may be a bit 'scary' to include social media in a book such as this, due to the rapidity of change, what may be considered the norm, and how all of that may change in a short period of time, social media are an important component of crisis communication.

Family support and victim assistance is a subject that could easily be a book in itself, but I have concentrated on the many *communication challenges* an organization faces when dealing with next-of-kin after a tragedy. And while many companies have done an excellent job, case studies show there is still a lot to learn, and the 'victim dimension' should never be neglected.

Leadership in a crisis is a subject covered extensively in case studies. How a government official or CEO speaks and what they focus on are quite often much more important than many in top management positions realize. Examples of this are highlighted from events ranging from a terror attack in Algeria to a train crash in Canada.

I have written this book to share what I consider to be vital lessons learned for handling the communication aspects of a crisis. My hope is that by reading this book you will be better prepared to deal with unforeseen events. We do not know what tomorrow will bring, but I hope the case studies and lessons learned that are detailed here will help you be better prepared to make the right decisions as you communicate in the minutes, hours, days, weeks after a tragic event occurs – and long afterwards.

I hope you find this book helpful, and that the insights and learning points I offer will help those who deal with crisis events and disasters to better serve the victims, their families, the public and the organizations they represent.

Kjell Brataas
Oslo, Norway

Acknowledgments

This book would not have been possible without help, assistance and inspiration from a number of people. I want to thank as many as possible in this chapter, but I apologize if someone who feels they should be mentioned is not included here.

First and foremost, I want to thank my immediate family. My wife Janelle has offered excellent support during the writing process, with practical advice regarding wording as well as emotional backing and encouragement. She suggested I spend a week away from home for writing purposes, and for many months she did not complain when I 'hid' in my office researching and writing this book. I am blessed with having a wonderful wife and life companion.

I also need to mention our three kids, Rebekka, Aleksander and Kristian. They too have been supportive and never complained when I was focusing on writing or could not attend an event at school. I am extremely proud of them and will love them forever.

The idea for this book was actually first put forward by one of my best friends, Ture Lundh, whom I have known since we both worked at Epcot in Disney World in 1988. When we were talking one day about my varied experience in the field of crisis communication, he suggested that I write a book on the subject. We discussed it further for the next days and weeks, and when I started writing, Ture provided invaluable help and support. He gave me concrete advice about structure and content, and most of the chapters in this book have been examined and corrected by Ture.

Someone else who has been of great help is Michael Kardos, whom I first met as a student in Austin, Texas, in 1993. He was my boss when I was an intern at the Texas Department of Commerce, and his guidance and advice on finding the right words for press releases and background articles have proven valuable for many years since. When I moved back to Norway in 1994, Michael and I kept in touch, and I am happy to say that he is still a good friend, who has contributed valuable advice for this book.

A venue that has been instrumental in the development of my crisis communication career is the World Conference on Disaster Management in Toronto. Its lectures and presentations have been top notch, but even more treasured are the professional connections and personal friendships I have made at WCDM. The people I have met in Toronto have inspired me in so many ways, and I have really enjoyed being a speaker and a member of the Advisory Board of the conference. I will not be able to name all the individuals I have met at WCDM, but I want to especially mention Adrian Gordon, Peter Power, Bob Jensen, Christopher Tarantino and Suzanne Bernier.

It was at the WCDM conference that I first met Australian Melanie Irons in 2014. She gave a fast-paced presentation about her volunteer work during the bushfire in Tasmania in 2013, and I immediately became intrigued by her story and her unique way of combining PhD studies, Facebook development and being a personal trainer. We talked for a while in Toronto, and I invited her to come to Norway to present at a KrisKom seminar I was arranging in Oslo the next year. Mel and her husband thus spent a week in Norway in May 2015, and we had a great time together, which was also spent discussing my book project. It was Mel who suggested I contact Taylor & Francis as a possible publisher, and I am very grateful for her advice, knowledge and friendship.

In 2005, the Norwegian government decided to set up a formal Emergency Support Unit within the Ministry of Justice. To make it happen, the ministry picked out five people with various backgrounds, who were given half a year to establish a system, write guidelines and properly equip the facilities two floors below ground at the Government Complex in Oslo. The head of the group was Per Brekke, and the other team members were Tore Drtina, Rolf Wegner, Stein Solberg and myself. The five of us worked very well together, and after the 'handover' of the unit to the Ministry of Justice, we continued to stay in touch. Every year since, we have met for a two days' seminar in a closed-down lighthouse in the Oslo fjord, where we update each other on personal matters and discuss subjects related to emergency management, disaster preparedness and crisis communication. We all look forward to the gathering, and I have learned much from these gentlemen that is included in this book.

Lindsay Crudele played an important role in handling the Boston Marathon bombings. I brought her to Norway for a seminar in 2015, and when I visited her in Boston in 2017 she presented me to a number of key people with valuable insight into crisis communication. Her introductions have been much appreciated.

I truly value my personal encounters with key people in the victim management profession, whose insights and advice are disbursed throughout the book. A few years ago I enjoyed a beer with Ken Jenkins when he visited Oslo, and his insight into the airline industry and its many accidents is probably equal to none. Another meeting that made a lasting impression was my talk with Heidi Snow in a hotel lobby in Boston. After experiencing a personal loss herself, Heidi set up an organization called ACCESS, which today is a world leader in providing peer support through 'grief mentors.' I believe many organizations and individuals have a lot to learn from her approach to grief counseling.

In June 2017 I attended a Victim Support Symposium in Dublin. The event was a fantastic arena for learning about victim support all over the world, and needless to say, I got to meet knowledgeable and interesting people with a variety of backgrounds and professions. The symposium followed the Chatham House Rules, which provide anonymity to speakers, so I am not in a position to name them here. Nevertheless, I want to thank the participants for openly sharing their personal experiences and lessons learned from disasters.

CHAPTER 1

Disasters in the Transportation Sector

INTRODUCTION

When a tragedy happens involving a train or a passenger jet, people all over the world pay attention. Often accompanied by pictures showing burning wreckage or distraught next-of-kin at an airport, these accidents touch the lives of all of us. Because flying today is such a routine task, we feel that 'It could have been me on that plane,' and we are again reminded of our own mortality.

The following case studies describe how crisis management, top leadership and communications teams handled two aviation disasters and one exceptionally deadly train crash. They each showcase the importance of social media, and each event is a reminder of how a CEO becomes an important figure – whether he or she likes it or not.

ASIANA AIRLINES: THE SPEED OF SOCIAL MEDIA

Case: An Asiana Airlines Boeing 777 from Seoul to San Francisco crashed into the runway while attempting to land at San Francisco Airport. Of the 291 passengers, 182 were injured and three lost their lives. The accident became a vivid example of how quickly news spreads through social media. Furthermore, it exemplified the importance of communicating well from all levels of an organization, and it proved that silence is not a recommended approach.

FIGURE 1.1 Asiana Airlines experienced several crisis communication challenges after flight OZ214 crashlanded in San Francisco. *Source*: NTSB.

The '15-seconds blog' called the accident 'Asiana's Crash Course in Bad PR' (15-seconds.com, 2013).
When: July 6, 2013

The timeline of the crash tells a compelling story of the speed of news spreading in the days of social media:

11.28: Asiana Airlines flight OZ214 crashes on runway 28 at San Francisco Airport.
11.29: Google employee Krista Seiden, who was at the airport ready to board another flight, uploads a picture of the crash on Twitter. Reporters from Sky News, CNN and NPR quickly reach out – through Twitter – to try to get her as a source for their news stories on the incident.
11.45: Survivors post photos of evacuation. One of them, David Eun, writes on Twitter (with an accompanying picture showing the crashed plane and evacuating survivors – with hand luggage): 'I just crash landed at SFO. Tail ripped off. Most everyone seems fine. I'm OK. Surreal ...'
12.08: First tweet from Boeing.
12.23: First tweet from NTSB (National Transportation and Safety Board.)
12.48: First tweet from SFO Airport – followed by 15 updates for the next 12 hours.
15.39: First tweet from Asiana Airlines.

By noon, there were more than 44,000 tweets about the accident, but the airline took several hours to post anything on social media about what had happened. Their excuse might have been that the Asiana headquarters are in Seoul, where it was night at the time of the accident and few people were in the office. For the first hours after the accident, therefore, their last Twitter message was from two days earlier – wishing everyone a happy Fourth of July.

> “
> Unfortunately, Asiana Airlines, with the world's eyes set on it, was slow to respond and was far from satisfying the insatiable need for more information in the hours after the crash.
>
> Shashank Nigam, *Simpliflying.com* (Nigam, 2013)

The way news spread on Twitter about the accident must be something close to a record, as a picture of the crash was published *one minute* after it happened. Neither traditional media nor a company's own communications team can ever compete with that, and this case study shows how hard it is to be on top of the news – and the importance of traditional and social media monitoring.

Asiana Airlines 'woke up' and published four updates on Twitter in the 12 hours following the crash. However, it took eight hours before they issued a press release, and when it came out it did not have any words on compassion or apology. Instead, it started with 'The following information has been confirmed' and continued to list the facts of the crash. It took nine hours for the phone number to the airline's next-of-kin hotline to appear on their website. Asiana chose not to engage in any conversations on Twitter, and they did not answer a single direct question through social media.

A Late and Silent CEO

You would think that when you run an airline, you would have no problem obtaining a ticket to fly to a crash site. Nevertheless, it took the president of Asiana Airlines, Yoon Young-doo, three days to arrive in San Francisco. What made matters worse was that he had nothing to say to the 50 or so reporters who were waiting for him at the airport. The CEO was quickly surrounded by journalists demanding answers, and the mob scene ended with Young-doo being chased back through the arrivals door. There was no other spokesperson in the U.S. who could talk on the company's behalf. This perplexing silence was repeated at a later press conference, where six of the 12 flight attendants from OZ214 took part. None of them spoke, and some hid their faces from the cameras.

There was also criticism in the media about how Asiana Airlines handled the surviving passengers. Many of them had no clothes, no luggage, no money and no passport, and several complained about not being heard or getting enough help. That they were instructed by the airline not to speak to reporters did not help matters (CBSNews, 2013).

Family assistance after an aircraft accident is taken seriously in the U.S., and federal laws protect the rights of surviving passengers and relatives. (See Chapter 9 for more information.) Asiana Airlines did not behave appropriately, and the company was later fined USD 500,000 for failing to help families after the July crash. This marked the very first time the Department of Transportation had issued a fine since a law about family assistance was introduced in 1997.

News from Authorities

American authorities and organizations displayed a very different approach to communicating about the accident. After the crash, the home page of SFO Airport had problems due to a very high number of visitors, but they made up for it with Facebook and Twitter as alternative platforms. They published constant updates on Twitter, made good use of hashtags and actively retweeted information from other sources. They were also able to answer individuals who had questions and comments about the situation at the airport.

NTSB – the National Transportation and Safety Board – sprang into action quickly. As is their goal, they posted a message on Twitter within one hour after the accident, and in the following hours and days they provided constant updates and streams of photos. During the first week of the accident, NTSB published 86 tweets with information on the crash, often providing factual information about the investigation and giving out pictures from the accident scene.

Victim of a Prank

If it hadn't had such a tragic backdrop, the prank played on the local TV station KTVU might just have been a tough lesson in the importance of verification. In their noon broadcast on July 12, the news anchor told viewers that 'KTVU has just learned the names of the four pilots who were on board,' then continued to read out four fake names of the pilots. It is a bit bewildering how the names made it all the way to the live broadcast, as at least one of the names – 'Captain Sum Ting Wong' – should have told TV producers that some fact-checking was in order. Soon afterwards KTVU made a public apology, but they still received heavy criticism for having allowed racist jokes. It did not help them, or NTSB, that a summer intern at NTSB had confirmed to the TV station that the names were correct. The intern was fired soon afterwards, and the news segment became an instant hit on YouTube.

GERMANWINGS: NO SURVIVORS, MANY QUESTIONS

Case: Germanwings flight 4U 9525 crashed into the mountainside at Massif des Trois-Eveches in the French Alps, killing all 150 on board. Several commentators said later that the crisis communication from Lufthansa (parent company of Germanwings) was a 'textbook response.'
When: March 24, 2015

Although the plane crashed in a remote area, news about the missing aircraft spread quickly. Flightradar24 was one of the first to report about the accident with a Twitter message that simply stated that the flight was lost from its radar. Within 90 minutes, the message was retweeted more than 2,000 times.

Germanwings published their first tweet about the incident within an hour and followed up with news on Facebook and their home page. In a later tweet they urged the public to monitor the website for updates, a task that became impossible, as the home page crashed and was unreachable for several hours. Social media therefore became extra important, and for the first 24 hours Germanwings published 24 messages on Twitter (14 in German, ten in English).

The CEO Steps In

Lufthansa, the parent company and owner of Germanwings, could have taken a sideline approach and let others handle the situation. That did not happen. Lufthansa's CEO, Karsten Spohr, played an important and visible role in the following days and weeks. His name appeared in the first Twitter messages from Lufthansa, thereby showing the world that the tragedy was being handled at the highest level of the company.

Speaking at the first press conference about the tragedy, Spohr said that the tragedy 'makes us speechless.' He continued to express shock and stated that the crash was 'our worst nightmare.' The CEO showed commitment and care, and a few hours later he recorded a video message in English and German that was published on YouTube. He said that Lufthansa would support next-of-kin in any way, anywhere in the world, and that 'safety in aviation is not a given.' There were several reactions to the YouTube video in its comments field, positive as well as negative. One person wrote that Lufthansa showed a personal side of the business and genuine feeling about the sad events. Nobody could take away the fact that at this most difficult of times, Lufthansa showed responsibility.

A Logo in Mourning

The two airlines involved in the crash quickly changed their logos and branding on social media, an action that has become the norm in the airline industry. Germanwings recolored its logo to a black

FIGURE 1.2 Carsten Spohr, the CEO of Lufthansa, appeared in a video on YouTube shortly after the Germanwings crash in the French Alps.
Source: YouTube.

and white version, and Lufthansa did the same, while also changing the cover page on its Facebook page to show only a black rectangle.

Lufthansa went even further and used the hashtag #indeepsorrow when it published news and comments on social media.

Family Support and Visits to the Crash Site

As described in Chapter 10 on family support, next-of-kin often want to see where their loved ones died. It was not possible, or advisable, to bring families of passengers and crew of the Germanwings flight exactly to the scene of the crash, but instead Lufthansa organized travels to a more central location. On March 26, 2015, two special flights brought family members from Dusseldorf and Barcelona to Marseille, where a Family Assistance Center was set up. More than 90 airline employees provided care and support at the center. Relatives who wanted to were taken to a location as close to the accident site as possible.

Books of condolences were made available at a variety of locations, and Lufthansa even set up a web page for digital condolences on www.indeepsorrow.com. The top of the web page states that it is 'A Place of Commemoration in memory of the victims of Flight 4U9525. This site collects thoughts and condolences for those who lost their lives in the tragic events of 24 March 2015.'

An aircraft accident creates emotions of sadness, disbelief and anger. Lufthansa and CEO Karsten Spohr made many good decisions and communicated well, but in the weeks following the crash, criticism arose regarding financial compensation. A single German school had been hit especially hard, as 18 of its pupils and teachers died in the crash. Their relatives launched several personal attacks on Spohr, claiming that the airline should have done more to help and that the CEO should have attended the funerals of their children. In several interviews, Lufthansa spokesman Andreas Bartels pointed out that Spohr had visited the crash site twice, that he had been to Barcelona and that he had attended memorial services in Haltern and Cologne. 'He could not speak with each of the more than a thousand relatives,' Bartels told reporters (*Australian Herald*, 2015).

In total, Lufthansa had more than 600 of its employees working on next-of kin support after the crash in the French Alps.

LAC-MÉGANTIC: CRISIS COMMUNICATION UNDERACHIEVEMENT

Case: A train with 72 tank cars carrying crude oil derailed in Lac-Mégantic, Quebec, resulting in a series of explosions that burned half of the downtown area and left 47 people dead. The communication efforts by the rail company and the behavior of the CEO were heavily criticized by locals, the media and PR consultants and have become a textbook example of how *not* to do crisis communication.
When: July 6, 2013.

FIGURE 1.3 At a speed of 65 miles an hour, a train with 72 tank cars derailed in the center of town of Lac-Mégantic. Sixty-three derailed tank cars were damaged, and a fire broke out almost immediately.
Source: Transportation Safety Board of Canada.

Of all the case studies in this book, the story from Lac-Mégantic is probably the one that describes the *least* effective crisis communication of all. Although the railroad company that owned the train involved was a large corporation (Montreal Maine & Atlantic), it did not seem to have any plans for crisis management or crisis communication. Its handling of the situation was so unprofessional that citizens of Lac-Mégantic felt neglected and forgotten, and the company received criticism in traditional as well as social media. It did not help that Montreal Maine & Atlantic had no corporate social media presence, and therefore no way of communicating quickly or interacting with journalists and citizens.

Much of the criticism after the crash focused on the head of the rail company, Ed Burkhardt. He decided to handle the crisis from his office in Chicago, and it took four days before he appeared in the town of Lac-Mégantic. He did not want to talk to the journalists who met him at the airport, and for a while he tried to avoid reporters, who swarmed around him on the street. Burkhardt seemed surprised at the anger that was directed at him, and when he finally decided to talk to the media, he did so at the side of the street. He had no prepared statements and no key messages, and in several of his comments he focused on his own situation. Burkhardt might have been trying to be funny, but his bad interaction with the press culminated when a reporter asked about his financial worth, and he answered: 'A whole lot less than I was Saturday' (Blatchford, 2013). In an interview with CNN, Burkhardt later explained that his actions had been misunderstood, pointing out what was obvious to everyone observing his news conference: 'Maybe I didn't present my case very well. But I'm not a communications professional. I'm a manager' (Coren, 2013).

> "People died. They died because MMA's train derailed and razed the downtown core of an historic Quebec town. Offer your condolences. Say you're heartbroken. For God's sake, show some humanity.
>
> Jonathan McLeod (McLeod, 2013)

Lac-Mégantic is a mostly French-speaking part of Canada, but Burkhardt only spoke English and did not bring an interpreter. He could easily have delivered a phrase or two in French to connect with the citizens, but he did not even convey 'Je suis désolé' – 'I am sorry.'

It was not only journalists who were handled inadequately and unprofessionally. Montreal Maine & Atlantic did issue a press release in French, but its translation from English was so bad that the words and sentences in the release became an insult to citizens who already felt neglected. Several suggested on social media that the company had used Google Translate, but Burkhardt later explained in an interview with the *Toronto Sun* that they had simply given the translating job to a member of their staff. 'That person's French was not very good. It was an embarrassment,' Burkhardt admitted to the reporter. Subsequent press releases were translated by a professional agency (Parent, 2013).

In his defense, it can be said that Burkhardt believed he could do his job best from the head office and not by traveling to Lac-Mégantic. He was also surprised by the hostile atmosphere when he finally arrived at the scene of the crash. However, the tragedy in Lac-Mégantic has become a telling example of how handling the communication aspect of a crisis – and top management involvement – is of utmost importance.

> "Burkhardt failed in his crisis leadership duties and as a result, the victims of the town of Lac-Mégantic were justifiably insulted and angry with him for doing so.
>
> Melissa Agnes, Crisis Management Strategist and Keynote Speaker. (Agnes, 2013)

BIBLIOGRAPHY

15-seconds.com. (2013, July 10). *15-seconds.com*. Retrieved from Asiana's Crash Course in Bad PR: www.15-seconds.com/asianas-crash-course-in-bad-pr

Agnes, M. (2013, July 11). *Melissaagnes.com*. Retrieved from http://melissaagnes.com/lac-megantic-crisis-a-look-at-rail-world-incs-crisis-communications-fail

Australian Herald. (2015, July 22). Families of Germanwings crash victims slam Lufthansa CEO's handling of disaster. *Australian Herald*. Retrieved from www.australianherald.com/news/235004731/families-of-germanwings-crash-victims-slam-lufthansa-ceo-handling-of-disaster

Blatchford, A. (2013, December 27). Head of railway at centre of Lac-Mégantic disaster: 'I was also a victim.' *The Globe and Mail*. Retrieved from www.theglobeandmail.com/report-on-business/head-of-railway-at-centre-of-lac-megantic-disaster-i-was-also-a-victim/article 16111290

CBSNews. (2013, July 17). Asiana crash victims: 'We didn't get any help' from airline. *cbsnews.com*. Retrieved from www.cbsnews.com/news/asiana-crash-victims-we-didnt-get-any-help-from-airline

Coren, A. (2013, July 13). Railway chairman: People in Canadian city 'wanted to throw stones at me.' *CNN*. Retrieved from http://edition.cnn.com/2013/07/12/world/americas/canada-runaway-train

McLeod, J. (2013, July 9). Lac-Mégantic rail disaster: Edward Burkhardt is a bit of a jerk. *Ordinary-gentlemen.com*. Retrieved from http://ordinary-gentlemen.com/2013/07/09/lac-megantic-rail-disaster-edward-burkhardt-is-a-bit-of-a-jerk

Nigam, S. (2013, July 9). Asiana Airlines crash crisis management 2.0 – case study and analysis. *Simpliflying.com*. Retrieved from http://simpliflying.com/2013/asiana-airlines-crash-crisis-management-sfo

Parent, M.-J. (2013, July 19). CEO Ed Burkhardt hopes Lac-Mégantic anger will one day 'dissipate.' *Toronto Sun*. Retrieved from www.torontosun.com/2013/07/18/ceo-ed-burkhardt-hopes-lac-megantic-anger-will-one-day-dissipate

CHAPTER 2

Natural Disasters

INTRODUCTION

When nature strikes, devastation and death tolls can be enormous. At the same time, crisis communication becomes extra important, as citizens in harm's way need detailed and correct information in order to survive.

This chapter focuses on flooding in Queensland and Calgary, a forest fire in Tasmania and a tsunami in Asia. They all teach us lessons about being prepared and using a variety of channels to reach the public. And as the story from Tasmania reveals, communicating with a large audience can now be accomplished by just one person dedicated to a page on Facebook.

The last case study in this chapter is a personal account of how the Norwegian government handled the tsunami that killed almost 230,000 people, 84 of whom came from Norway. I worked on it from the government side, and as the case study reveals, we were not prepared ...

FLOOD IN QUEENSLAND

Case: A massive flood and tropical cyclone swept through Queensland, Australia, affecting more than 90% of the state and millions of people. The Queensland Police Service (QPS) used social media in an effective way and experienced an increase in

followers on Facebook from about 17,000 to more than 100,000 in 24 hours.
When: December 2010/January 2011

Only half a year before the flood occurred, the QPS had started experimenting with social media as an effective way of reaching the public. Interest in their social media channel grew steadily through word of mouth and by active promotion of the channels in press releases and on web pages. By November 2010, the QPS had 8,000 'likes' on Facebook and 1,000 followers on Twitter.

The QPS were therefore in good shape to start informing the public about dangers and give directions about what to do when, in December 2010, bad weather set in. Their first posts on social media were copies from press releases, and interest and demand grew quickly. After only two weeks, the number of 'likes' on their Facebook page escalated from about 8,000 to 17,000.

Still, this growth did not amount to a lot compared with what happened after January 10, 2011. On that day, a significant flash flood occurred in south-east Queensland, and the number of 'likes' on the QPS Facebook page increased from about 17,000 to more than 100,000 – in 24 hours (Queensland Police Service, 2011).

The Use of Facebook – a Success Story

An important reason for the popularity of their Facebook page was that its content was up to date, an achievement in itself for a public office used to having rigid clearance routines for information issued in press releases. The communications team were trusted to use their judgment, and journalists soon started seeing the QPS social media channels as their main provider of official information. Traditional media did a good job in promoting the channels, often citing the Facebook or Twitter pages and thereby informing the public of where they could find information directly from the source.

Another cause for success was the capability of the social media team to respond adequately and in a timely way to questions coming in on their Facebook page. Although its popularity and usage had grown immensely, the QPS media team attempted to pick up on trending questions and comment where needed, something the public valued, as they experienced concrete proof of being 'heard' and taken seriously at a difficult time.

Twitter: More than Short Messages

The QPS also made extensive use of Twitter. In addition to putting out messages and updates, the police used the channel for live-tweeting vital information from briefings and press meetings, including new facts and advice to the public. All tweets carried the hashtag #qldfloods, which made them easy to follow. The consistent use of the hashtag also made the public want to retweet, as they knew these messages were of great importance.

In any disaster, rumors and misinformation spread quickly. To counter this, the QPS coined the hashtag #mythbuster and used it extensively in tweets that killed rumors and misreporting. In addition to reaching the public, the stream of #mythbuster tweets became important for journalists to follow so that they could report correctly about the situation.

The QPS Media Unit received several positive comments about their use of Twitter during the flood; here are a couple of examples:

> @QPSMedia is doing a stellar job in a real crisis. Thank you QLD Police Service. #qldfloods
>
> @QPSMedia Thank you for using Twitter to keep everyone up to date. And for clearing up all the misinformation so quickly. #qldfloods

World Records?

- The QPS Facebook page is said to be the largest police Facebook page in the English-speaking world. It has some 800,000 followers – in a state with a population of 4.5 million.
- The QPS Facebook page had 39 million views in one 24-hour period.

Extra Eyes and Ears through Social Media

During the flood in Queensland, the local police learned the value of social media in a variety of ways. Not only were they an excellent tool for providing information *to* the public and the media, but they also became a useful means for the public to report dangerous situations or vital facts the police needed to know about. For each new 'like,' the QPS got an extra eye on the ground, thereby also improving situational awareness and their ability to handle difficult circumstances.

> **On Social Media from 2010**
>
> The QPS began using Facebook, Twitter and YouTube in May 2010 (Queensland Police Service, 2011).
>
> Their aim was to:
>
> - Claim social media presence
> - Engage in a two-way conversation between the QPS and the public
> - Develop an online community of followers before a disaster occurred

TASSIE FIRE: ONE PERSON COULD HELP

Case: A forest fire in Tasmania, Australia, left hundreds of people in need of support. Melanie Irons, a local PhD student, set up a Facebook page that gave vital information and also mobilized people to help others.
When: The fire started on January 4, 2013.

As news about the fire spread, Melanie Irons was at her home in Hobart preparing her PhD. She immediately realized that many Tasmanians would be needing help, and for a while she pondered what she could do. Make sandwiches? Offer to take care of people's animals? 'I wanted to do more, and when I saw a variety of posts on Facebook asking for or offering help, I realized I could set up a Facebook page that could facilitate this need,' Irons said at a conference in Oslo in 2015. The result was: 'Tassie Fires – We Can Help.'

At first, the page received only two likes and two comments. To help boost awareness, she decided to call the local ABC station and told them about her project, and they agreed to put the word out. After only one and a half hours, 3,000 people were following the page – after 24 hours, it had reached 17,000 followers. During the first hours, she posted lots of requests for help combined with messages from people who wanted to help. She also started copying news releases from the government, which she published on her Facebook page. When she later analyzed all its contents, she found that if she simply copied the text from the government, the posts would get maybe 30 comments, likes and shares. When she converted the language and made it more casual and accessible, the post would receive hundreds more comments, likes and shares.

FIGURE 2.1 'My name is Mel …' The Facebook page 'Tassie Fires – We Can Help' featured a personal introduction from its creator Melanie Irons.
Source: Facebook screen shot.

An Impressive Reach

Thanks to Irons' hard work and dedication, 'Tassie Fires' became *the* information hub for people needing assistance or offering help. On average, Irons posted a message every 4.3 minutes for two weeks, reaching a total of more than 3,000 posts in the same period. The page had more than 20,000 followers and nearly 3 million 'impressions' during the first two weeks of its existence.

One Individual

The Bushfire Inquiry was published by authorities in 2013. Its nearly 300 pages refer to numerous roles, but include the name of only one individual – Melanie Irons (Tasmanian Government, 2013).

The posts varied greatly in content, length and tone. There were many requests for essentials such as water or gas, and farmers also needed help with their livestock or transportation. At the same time, people were volunteering (offering anything from reptile handling to babysitting), and Irons worked closely with the Red Cross.

The fire spread quickly, and for many Tasmanians, their only option was to run for the water. At some places even the beach was on fire, so the situation was definitely life threatening, with locals and tourists trapped in several locations.

Facebook vs. Twitter

Irons also started cooperating with the police and government officials. In Tasmania, the government had decided that Twitter would be their main social media channel (Tasmania Police were developing a Facebook page), but the Tassie Fires page showed that the public was on Facebook. On several occasions the government called Irons and asked her to post their information on her Facebook page, and at one point people from the government came to her house to discuss strategies for reaching as many people as possible through the Tassie Fire Facebook page. A clear learning point is, therefore, that it is important to know which platform is the best to reach your audience – and if in doubt, use a scattergun approach and as many different platforms as your resources allow.

In addition to providing a digital space for requests and offers of help and practical support, Tassie Fires – We Can Help became a forum for people to tell their stories – a critical function for improving psychological recovery. It also provided fundraising information, and Irons took an active role in making sure donations and offers of help were handled quickly. 'I wanted to "suck dry" the volunteers while they were still interested!' Irons said at the KrisKom conference in Oslo. She also pointed out that it was important for people to see the results of giving, which in turn made even more people want to help.

> "
> Engagement with the page continued to soar and Mel was heard on radio, seen in the newspapers and acknowledged by high profile people such as the Tasmanian Premier. This generated even greater awareness about the page and what Mel was trying to do. From the very beginning, Mel placed her personal contact information on the page, which she believes gave the site credibility. People emailed or called her if there were any issues needing to be followed up.
>
> From the official inquiry. (Tasmanian Government, 2013)

Although she was alone in the beginning, Irons did not do all the work on the Facebook page herself. She got support from 'digital volunteers' who helped moderate the posts and follow up on those messages that needed immediate attention. 'I got 10 people to help me; so far I have only met one of them face-to-face,' Irons said in Oslo. The Facebook page thus showed that people can help in a variety of ways, and that the concept of a VOST – Virtual Operations Support Team – shows that you can help in an emergency even by sitting in front of a computer.

FIGURE 2.2 Since the fire, Dr Melanie Irons has become a sought-after presenter and has given talks internationally, including in Toronto, Washington DC, Frankfurt, London, Belfast, Edinburgh, Johannesburg, Wellington, Riga, and Oslo. She finished her PhD in 2015.
Source: Kjell Brataas.

Irons got lots of positive feedback for her volunteer work during and after the fire. She received numerous Facebook notes and email messages praising the initiative, such as these:

- The page was the right page at the right time. We were devastated hearing about the fires, and you provided a means for all of us to really help. The first week was absolutely awesome. I cried lots with the fantastic things people were able to do. Not something I'm prone to do much of – I'm male!! Your page kept us all in touch with real time needs and we really felt what we were doing was actually making a difference.
- When we were isolated on the Peninsula, this page was the life-saver of my emotional wellbeing. Just reading of the goodness of strangers and how every one was helping took my mind off the fear I felt of being so isolated.

Advice and Lessons Learned

Some of the advice Irons presents can be summarized in these points:

- Utilize social media to give your organization a 'human face.'
- Use volunteers. Actually use them.
- Have a rumor plan.
- Have a plan for keeping records.
- Sign off at night – don't forget to sleep.

CALGARY FLOOD: CRISIS COMMUNICATION AT ITS BEST

Case: A flood due to increased water levels in the Bow and Elbow rivers caused the evacuation of 32 communities. Consequently, up to 80,000 citizens had to temporarily leave their homes. The City of Calgary and other government agencies handled the communication aspect of the crisis in a highly satisfactory manner, especially by using social media and by having an active mayor.
When: June 2013

The flood in Calgary turned out to be Canada's costliest natural disaster, and normal activity in the city was suspended for almost a week due to damages caused by the flooding. The City Hall and the Municipal Complex were damaged, but fortunately

FIGURE 2.3 According to the City of Calgary's web page, 'Flooding can occur at any time with little to no warning.' That was the case in 2013.
Source: Wilson Hui.

the city had invested heavily in a state-of-the art operations center, which proved vital for handling the many tasks of crisis management, clean-up and recovery. Another important asset was Mayor Naheed Nenshi, who saw it as one of his key functions to inform the public.

It is rare that government offices receive positive feedback for the way they communicate during a crisis. However, that was the case for the City of Calgary and Calgary Emergency Management Agency (CEMA). According to a report by the Conference Board of Canada – *Forewarned and forearmed: The Calgary Emergency Management Agency and the 2013 flood* – the city's quick action was 'lauded by the public,' and their use of social media meant that city officials could 'lead the conversation from the start.' In fact, 73% of Calgary's citizens strongly agreed that the city had communicated effectively with them during the crisis (Vroegop, 2014).

In this chapter, we will look at what contributed to the success of the communication efforts:

- A modern Emergency Operations Center (EOC) served as a hub for communication, control and collaboration.

- Government officials and offices focused on openness. Media was allowed into the EOC, and journalists were given police radios to be able to follow – live – what was happening.
- Social media were used extensively; the hashtag #yycflood showed up an average of 32 times every minute over a 10-day period.
- The mayor did his utmost to inform the public and inspire and applaud his people.

The EOC

The necessity for a modern EOC became apparent after previous flooding in Calgary in 2005. Coordination was somewhat lacking, and it was therefore determined that the city needed a facility from where all services could operate and coordinate the responses to an event. Construction was started in 2009, and the operations center opened in October 2012.

FIGURE 2.4 Calgary's Emergency Operations Center has dedicated facilities for journalists.
Source: Calgary Emergency Management Agency.

Thus, the City of Calgary had modern facilities to operate from during the flood in 2013. The center is big enough to host a variety of agencies and collaborators, and it has its own media center – which made it possible for journalists to be close to where decisions were made. Inviting media to the core of a crisis operations center might seem daunting, but the Calgary flood proved that working with the media and having them close by can make crisis communication easier and more effective. Another aspect of the size of the EOC was that it was possible to have breaks and organize media briefings and on-the-spot consultations in the same place, which in turn meant less time organizing and traveling to meetings and more effective and less stressed personnel, collaborators and journalists.

> CEMA coordinated the city-wide efforts of evacuation, recovery and communication during the flooding. The EOC is home to CEMA on a daily basis. The facility also contains a back-up 911 call center and a city data center.

Social Media and 'Twitter Jail'

When flooding started in June 2013, the City of Calgary was no newcomer to social media. Its Twitter account @CityofCalgary had more followers than any other Canadian city, and earlier (although smaller) events had been utilized to test the use of social media for two-way communication with the public.

Communications specialists within the city quickly realized the importance of a hashtag, and #yycflood became a significant tagline to communicate about the event on social media. Twitter was the main social media channel utilized by the City of Calgary and the Calgary Police Service. Other services included Facebook, YouTube and Flickr.

The Calgary Police acted professionally and actively on Twitter. Not only did they respond to questions and comments; they also used the channel to create a strong bond between those who worked on mitigating the flood and the people it

FIGURE 2.5 The Calgary Police did an excellent job of communicating through Twitter.
Source: Screen clip from Twitter.

affected. One example is the response they posted on Twitter to the question 'What do we do about vagrants and questionable people roaming the streets in suburban communities because of the flood situation?' Their reply: 'Suspicious people can be reported by calling the police. Homeless people may appreciate a sandwich.'

Part of the success with social media can be attributed to the fact that management levels and individuals in high-ranking positions had a positive attitude towards the use of social media. Many communication efforts and messages can be hindered by red tape and approval processes, but this was not the case in Calgary. In the report from the Conference Board, the Mayor's Office as well as CEMA senior leadership are credited with a 'push for the development of a city-wide corporate social media capacity' (Vroegop, 2014).

In fact, the Calgary Police used their Twitter channel (@CalgaryPolice) to such an extent that on the evening of June 20 it found itself in 'Twitter jail.' Their office had sent out such a high number of tweets – and replied to many incoming messages – that automatic systems at Twitter deemed the activity to be a sign of something wrong. As a result, the Twitter account of the Calgary Police was shut down. Obviously, this resulted in hectic activity trying to get the account restored as quickly as possible. Several individuals contributed, including a platform relations manager for Twitter. Only 40 minutes after it was blocked, @CalgaryPolice was back up and working again. Reinstating its professional manner in dealing with social media, the very first 'new' message from the police thanked Twitter for getting the account back up, followed by the short message 'Testing. Are we back?' and a back-to-normality message saying: 'Update. Our account has been restored. Sorry for the delay in responding to all your #yycflood inquiries. #yyc.'

Although social media proved extremely valuable as a communication tool, other channels were also in use. Radio and television interviews with crisis managers appeared regularly, and a blog as well as two free apps from the City of Calgary also helped get messages across.

The Role of the Mayor

During a crisis, people affected want to see the face of the man or woman in charge. Mayor Naheed Nenshi was this person, and his outgoing personality, humor and long hours undoubtedly contributed to the success of the city's crisis communication. Nenshi gave a number of press briefings and many more interviews, and he made several visits to various sites in the city where flooding had occurred. (Most of them took place during the night, so as not to disturb clean-up crews.)

FIGURE 2.6 The mayor of Calgary used every opportunity to praise his staff, who worked hard on handling the flood.
Source: CBC LICENSING.

Many politicians are capable of handling the media, but what made Nenshi stand out was his constant focus on praising his staff. He used every chance he had to talk positively about the police force and fire officials, and when asked on TV by reporter Ian Hanomansing from 'The National': 'What is the message for the country?' Nenshi had the following answer: 'First of all, the message is: Hug a public servant.' Nenshi then went on to commend all those who worked on the crisis and the dedication they displayed ('The National,' 2013).

Nenshi was also active on Twitter, and on many occasions, he used his Twitter account to reinforce messages from the police and from the City of Calgary.

In an interview with CBC, Nenshi explained that he had three main tasks in managing the flood (Rieti, 2013):

1) Give people the information they need to stay safe.
2) Give hope and courage to people affected by the floods.
3) Stay out of the way as relief efforts continue.

EOC Underground Facilities

Most of the EOC building is underground – in part because of security, but also because the building is designed to blend into the surrounding area rather than compete with it. Extensive community consultation was completed prior to building construction in 2008.

The EOC has built-in redundancy and is capable of running 'off the grid' for a minimum of 72 hours. Several generators power the EOC in the event of a grid interruption.

The EOC cost approximately $54 million to build.

High River – Lessons Learned from an Appropriately Named Town …

The community of High River south of Calgary was one of the hardest hit by the flood in 2013. As the entire town was evacuated, crisis communication became essential. Joan Botkin, who was the

Communications Manager for High River at the time, points out these lessons learned:

- Implement an early notification system.

After the flood High River implemented a system called High River Alerts that enabled us to issue alerts through a direct message to anyone who subscribes via phone (voice), text, or/and email. Residents choose how they want to be contacted. This system has now expanded and has become a regional network for the area with eight communities participating. The name has changed to SCAN (Safe Communities Alert Network).

- Develop a comprehensive crisis communications plan.

It should have a clear description of how supporting agencies will work with the local communications department. It should also provide clear descriptions of the roles and responsibilities for each member of the communications team. Creating it under the ICS model (Incident Command System) allows the organizational chart to shrink and expand depending on the scope of the crisis. As well, the plan should include back-up personnel within the organization who is capable of fulfilling the role, although their job may not be communications.

- Create a regional crisis communications support team.

Most municipal communications teams are very small (ours was an office of two), and this does not provide the capacity to respond to a crisis effectively. Also, it allows no breaks for the information team. By creating partnerships with other towns and regions, you can double the size of the communication response. In most cases the partners can provide support from their desks through social media and the web.

- Save stuff in the cloud.

During the first 36 hours we lost all forms of communication including Internet, land lines, cell phones and didn't even have a printer. Our website also crashed for part of the time due to the heavy traffic. The Town's servers were down for several days and we were lucky that we had set up a separate Gmail account prior to the flood since the Town's service was not working.

- Appreciate the importance of paper.

Everyone always lauds the benefits of social media, but when you're dealing with people in extreme stress, like we were, they don't retain things. Providing printed materials is comforting and lets them reread the message as many times as they need to. Essentially – use as many different tools as possible to reach as many people as possible. The entire town of High River (12,900 residents) was evacuated and not allowed back for almost a week. Some weren't allowed back for several weeks. Everyone was scattered all over the place so we needed to think of as many different ways to reach them as possible.

Repurposing existing online tools like an e-newsletter that are considered trustworthy and credible by residents can help organizations in sharing information.

We had a regular e-newsletter that we repurposed into an information bulletin that we sent out several times a day during the initial hours of the crisis. Subscription to the site tripled within 48 hours. It remained one of the key information tools throughout the crisis and recovery.

THE TSUNAMI: A WAVE OF CHALLENGES

Case: An earthquake measuring 9.3 on the Richter scale triggered a large sea wave that struck the coast of Thailand, Sri Lanka, India and ten other countries. The waves killed more than 230,000 people, including 33 Americans, 151 British, 543 Swedes and 84 Norwegians (26 of the Norwegians being younger than 18). Crisis communication, especially the first days after the tragedy, was chaotic. Governments and their respective agencies were not up to standards and were duly criticized for not being prepared for handling the situation.
When: December 26, 2004

December 26, 2004 was supposed to be one of the happiest days in their lives. Thousands of tourists had chosen to spend their

Christmas 2004 vacation in Asia, many of them in resorts along the western coast of Thailand. Khao Lak was especially popular; with accommodation ranging from first-class hotels to a variety of beach bungalows, it was ideal for families with children.

The morning started like any other day in 'paradise,' with lazy breakfasts, beautiful sunshine and happy travelers. However, beachgoers noticed something strange – the sea water was receding. This made people curious, and many wandered into the sea to check out what was happening and to take pictures of the fish jumping in the sand. Yet, most people along the beach had no idea that something was about to happen – and that they would soon be in a life-threatening situation.

At around 10.30 in the morning, an enormous wave hit the western coast of Thailand. Where the beach was at its most shallow, the wave grew tallest, meaning that the child-friendly beach of Khao Lak quickly became a disaster zone. The waves reached 10 meters (33 feet) in height, crashing into everything in their path and instantly making paradise into hell. Most of the guests in the hotels had no warning whatsoever of what was happening, and they had no time to reach for their kids or try to find safety. Soon, everything, everybody – men, women, kids – furniture, minibars and cars were floating around in muddy, swirling water; survivors later described it as being swept up in a whirlwind of terror.

Coordination and Communication Challenges

At the Norwegian Ministry of Foreign Affairs in Oslo, a duty officer was notified of the tsunami at around 5 am Norwegian time. He then contacted Norway's embassy in Bangkok and told them about the news. Soon afterwards, the switchboard at the ministry started receiving phone calls; their numbers soon escalated in such a way that it became difficult to make outside calls through landlines.

Two challenges became apparent right away: how to find out what had happened to the 4,000 or so Norwegians vacationing in Thailand, and how to get survivors back to Norway. Government officials soon started cooperating with various travel agencies while at the same time sending two extra people to staff the embassy in Bangkok.

Information about what had happened and how many were affected took several days to work out accurately. On the evening of December 26, Reuters reported that a total of 11,300 individuals had died in Asia; two days later, they reported the number to be 26,000. At the same time, Norwegian newspapers printed dramatic pictures and stories from survivors, and Norway's national TV station NRK even had a reporter near Phuket who had been vacationing there with his family and who reported live through a bad telephone line to viewers in Norway. All media outlets in Norway told their journalists on vacation to report to work immediately, and TV and radio stations had several extra news broadcasts already from the first day of the tragedy.

Telephone Support, with Not Much Support …

In 2004, I was a communications advisor with the Ministry of Health and also part of a support unit that was to be called in when there was a need for extra personnel. On the morning of December 27, I made contact with the Communications Unit at the Ministry of Foreign Affairs, offering my assistance. They declined, saying 'We can handle it.'

With so much uncertainty – understandably – families in Norway were desperate to find out what had happened to their loved ones in Thailand. The Ministry of Foreign Affairs received hundreds of phone calls, and their numbers increased dramatically in the afternoon of December 27 when a hotline number was published and news about 13 dead Norwegians got out.

I *then* got a phone call, urging me to come to the Ministry of Foreign Affairs as soon as possible to assist with answering the hotline. Arriving at their office in the afternoon, I met colleagues from other ministries – and a room and equipment that were not at all designed for handling a large number of incoming calls. About ten of us were asked to sit around a large table, where 12 telephones (of the normal kind with no headset) were constantly ringing. Beside each was a stack of papers; no forms, no instructions as to what to say or do. We were only told: 'You can now start answering.' We did.

At the other end were all kinds of callers with a variety of questions and information, many in tears and desperate. Most were wondering if we knew anything about their family member, and gave us their names, height, eye color and so on. Some Norwegians also called from Thailand, informing us that they were in good shape or that they needed assistance with evacuating. We wrote all this information down, one piece of paper for each phone call. The papers were then piled on top of each other at a central location in the room, but as hours passed we noticed that nothing happened to the stack of papers. When later in the evening I got a phone call from a person I had talked to hours earlier, asking if there was any news about his missing person, I could therefore glance at the pile of papers and know that there was no news to give out.

To lessen the burden on the telephone line, the Ministry posted a generic email address to the public. In only a few hours, more than a thousand emails came in, but unfortunately there was no system in place to handle all the incoming questions. Another obstacle was what kind of database to choose. During the first two days of the tragedy, Word, Excel and Access were tried, before agreeing on File Maker Pro. Merging the information into one system took 20 people several hours, and by the evening of December 29, the Ministry of Foreign Affairs had 8,000 registrations in File Maker Pro.

Telephone enquiries grew to a new level on December 28, when the Ministry of Foreign Affairs issued a press release stating that 700 to 800 Norwegians were missing. Two days later, Norway's prime minister, Kjell Magne Bondevik, said that 460 were missing and that the country had to realize that many of them were dead. On December 31, Norway's largest newspaper *VG* printed the following headline, covering the whole of its front page: 'At least 100 Norwegian children might have died.'

Criticism

The media soon started reporting about the lack of coordination and support from the Ministry of Foreign Affairs. Survivors in Thailand complained that they could not get hold of anyone from the embassy in Bangkok, while family members in Norway criticized long waits and few updates from the hotline at the ministry in Oslo.

Desperate, and in need of information about their loved ones, several relatives of missing Norwegians jumped on a plane to Thailand to take part in the search effort themselves. Having experienced the chaotic situation at the telephone hotline, they got even more frustrated when they had problems finding *anyone* from the Norwegian authorities in Bangkok, Phuket or Khao Lak. The frustration culminated at a hastily organized meeting in Phuket, where the Norwegian ambassador to Bangkok clashed with an individual representing the 30 or so relatives at the meeting. There was so much heated debate and discussion that at one time they had to call a 'time-out' and then resume the meeting when everyone had had time to calm down. Several reporters watched what happened, and as a result Norwegian newspapers the next day carried several stories about frustrated relatives and incompetent authorities. One headline read 'Strong criticism against the government'; another simply stated – in bold letters – that there was 'Total confusion.'

Airport Support

News reports showed that many of the survivors were in bad shape, skimpily clad and in shock. They had been through a traumatic situation, they had only narrowly survived, and many of them still did not know whether their family members were alive, missing or deceased. At the same time, journalists had a great 'need' to interview the returning survivors, and in order to protect them from a media frenzy, the Ministry of Health decided that extra staff and communications experts would assist at the airports in Norway and Sweden.

Evacuation processes in Thailand were chaotic. Some survivors flew home on their scheduled flights, but governments in several countries also sent chartered planes and special Medevac planes that could accommodate several individuals on stretchers. At Phuket Airport, survivors were desperate to get away, and they often got a plane ticket to a city that was not their final destination. Many Norwegians therefore ended up on a plane going to Gothenburg or Stockholm in Sweden.

On December 30, I was asked to fly to Arlanda Airport in Stockholm to assist Norwegian survivors. Upon arrival, I was positively struck by the resources and care displayed by Swedish authorities at the airport. The Swedish government also got lots of criticism for their handling of the tsunami, but they deserve praise for the way they set up survival support at Arlanda. At a designated area of the airport, special fences had been put up so that those arriving from Thailand could walk into the arrival hall without anyone seeing them from outside. This meant that they were protected from curious travelers and journalists, and that they could feel safe and taken care of. Psychologists and staff from local authorities were on hand to answer questions, and food and drink were provided. The McDonald's restaurant at the airport constantly brought fresh hamburgers, and IKEA had donated a big pile of teddy bears for the arriving children.

Many of those arriving wanted to talk. I therefore spent several hours hearing the horrific stories from the survivors, and being a father of three, I paid special attention to those arriving with small children. Our second task was to get Norwegians on to their next flight, which people from SAS and Star Tours handled exceptionally well. Another group that did a great job was representatives from the Norwegian Church Abroad. With their experience and insight into grief and counseling, they were a great resource for survivors who wanted to talk or a shoulder to cry on.

I returned home in the afternoon of December 31. There had been no sleep at Arlanda, and I was therefore physically and mentally drained and slept all the way on the plane back to Oslo. Needless to say, we cancelled our New Year's Eve celebration that night.

Direct Communication

At the beginning of January 2005, survivors and next-of-kin formed a national support group. They had several objectives; perhaps the most important was to be a central point for contact with the government and other authorities handling the tsunami and its aftermath. In order to facilitate this in an effective manner, it was decided that there should be frequent meetings between the support group and authorities. It turned out to be a good choice, and although members of the support group were highly critical and vocal, they valued the chance to talk directly to those in charge.

They had many questions. They wanted to know what was being done to find survivors, what the Norwegian government was doing to assist local authorities, whether Norway could send 'sniffing dogs' to the beaches in Thailand to help locate the deceased, and how the dead would be transported to Norway. The meetings (I attended all of them) were quite emotional, and on several occasions representatives at the meeting – from both 'sides' – shed a tear or two when hearing about the ordeals in Thailand and the frustration and anger many felt when returning home.

On 'our' side of the table were representatives from several ministries. I was there from the Ministry of Health; other government officials came from the Ministry of Foreign Affairs, the Ministry of Justice, the Directorate of Health, the police, psychologists, the Norwegian Church Abroad and other experts. Often they could answer questions from the support group right away, but some questions needed more looking into and were therefore answered at the meeting the following week.

Back to where It Happened

It was during one of these meetings that the idea of taking next-of-kin on a trip to Thailand was first discussed. Relatives of those who died or were missing had a number of questions about the circumstances surrounding the tragedy, and some were wondering 'Why didn't they swim away?' or 'How come he couldn't save his son?' The support group actively reinforced the idea, and in early March the Ministry of Foreign Affairs started looking into the feasibility of a trip. Obviously, there were uncertainties, but at the same time those involved envisioned that a visit 'back to where it happened' could have many merits.

I personally think that all the criticism and the bad light authorities found themselves in played a role in the decision to go ahead, maybe hoping to finally get some positive feedback from next-of-kin and the media.

Whatever the underlying cause, the Norwegian government decided to arrange for family members of those who had died in Thailand and Sri Lanka to travel to the area where their loved ones had perished. Families of the four who died in Sri Lanka were given financial aid for traveling by themselves, but for people headed for Thailand there had to be a more formal set-up, which included flights to Phuket, arrangements for personal visits to the disaster areas, and a memorial service with speeches and music.

To arrange for all this, the Norwegian Ministry of Foreign Affairs established a small task force that was in charge of logistics, support group cooperation, media handling and communicating with the next-of-kin who would be traveling. I was part of the group, and as can be imagined, we realized in one of the first meetings that we had a challenging job ahead of us – and many open questions:

- How many will be traveling?
- When should the trip take place?
- How do we travel?
- Who should be eligible?
- How many from each family should be allowed to go?
- What would be a good location for a memorial service?
- Which hotel would be suitable?
- How do we inform group members about travel plans, changes and itineraries during the trip?
- What kind of media interest will there be?
- What costs should be covered?

It soon became apparent that we could not find the answers or plan the details of a complicated trip without seeing the destination firsthand. The task force therefore traveled to Phuket in Thailand at the beginning of April 2005 to survey possible hotels and transportation options and to meet with local authorities and travel agents who could help us with the logistical challenges at hand.

One of the first meetings we had upon arrival in Thailand was with Espen Westlie, who was head of the Star Tour agency, which had been responsible for most of the Scandinavians traveling in Thailand at the time of the tsunami. Westlie became a vital asset in planning for the trip, and his local contacts and geographical knowledge were invaluable for the rest of us, who had not been to the area before. Other helpful advice came from ministers of the Norwegian Church Abroad who had been assisting victims in the days after the tsunami, and representatives from the Norwegian support group, who had opened a small office near Phuket.

The 80 Norwegians who died in Thailand had been to various parts of the country when tragedy hit. A few had stayed on the island of Phi Phi, some had vacationed at Krabi or Phuket, while most of them had spent the Christmas days of 2004 near Khao Lak. From talking to the support group, we knew that each next-of-kin who would be traveling wanted to get as close as possible to where their loved one had died. This meant that the group would be spread widely, and that we had to arrange for various transportation options. For the Phi Phi families we chartered a large yacht, but the trips to Khao Lak proved the greatest challenge. Looking at maps of the area in meetings in Norway, we originally thought we could rent several buses that could bring our guests to Khao Lak as a large group. However, when we got to the beaches of Khao Lak ourselves in April 2005, we soon realized we had to reconsider. The resort area was much more spread out than we had thought, and the Norwegians who died there had stayed at hotels many kilometers apart. Bringing everyone there as a group was therefore not possible, as they would need individual transport modes to 'their' part of the beach. Alas, walking was out of the question too – the age of some of the travelers, combined with a tropical climate, meant that we could not risk having our guests wander around in the blistering heat.

Another challenge was finding the right hotel. We knew we would be a large group, so smaller boutique hotels or romantic bungalows would not be possible. At the same time, we were told that the next-of-kin on the trip did not want to stay too close to the beach (but they would like the beach to be within walking distance), and they would not accept staying on the first floor. While doing research in Thailand, our task force therefore spent considerable time scouting suitable accommodations.

Next, we needed to find a close to ideal place for the memorial service. We looked at everything from school yards to a mountain lookout, but we soon decided the location had to be a beach. Again, there were all kinds of considerations, including the need for shelter in case of rain, access to bathrooms, technical facilities for microphones and speakers, and food and drinks for guests and staff.

We spent almost two weeks in Thailand for planning purposes, and when we returned we had a pretty good idea of how the trip should be organized. By then, it was also clear that there had to be two different trips, one in May and one in October. As not all the 84 Norwegian casualties of the tsunami had been identified, several family members did not want to go to Phuket when the status of their loved one was still unclear.

In the afternoon of May 2, 2005, a chartered plane took off from the airport in Oslo with 200 next-of-kin on board. Other passengers included support staff, representatives from the Norwegian Church Abroad, family liaisons from the police, and a variety of doctors, nurses and psychotherapists. They arrived the next morning (Phuket time) and were transported by bus to the Royal Orchid Resort in Karon. The members of the task force had arrived a few days earlier, and we met the travelers at the hotel, where we handed out a short note explaining the hotel's amenities and location and an invitation to an informal meeting the same afternoon. At the meeting, we explained the agenda for the next days, gave advice about coping with the heat, and also encouraged

FIGURE 2.7 The beach of the Katathani Hotel was chosen as the site for the Norwegian memorial service to commemorate victims of the tsunami in Asia.
Source: Kjell Brataas.

FIGURE 2.8 When Norwegian next-of-kin of tsunami victims visited Khao Lak in May 2005, Hilde Sirnes (left) from the Norwegian Church Abroad and Kjell Brataas from the Ministry of Health offered support and practical advice.
Source: Cathrine Andersen.

participants to undertake an activity not related to the tragedy, such as riding an elephant or trying scuba diving.

The next two days were set aside for family members to visit the area where their loved ones had died. For each family wanting to see Khao Lak, we had rented a mini-van and a driver who had been told which hotel their next-of-kin had stayed at. Each family could make use of the mini-van however they liked. In the mornings, we had arranged for locals to set up a flower stand in the hotel lobby, and most families bought beautiful bouquets, which later in the day they would place in the sand on the beach. We also rented several large buses that drove to Khao Lak early in the morning. On board were support teams of ministry officials, police and healthcare providers, and we also packed bags of ice, soft drinks and light refreshments. Upon arrival in Khao Lak, the buses were parked in various predetermined locations along the beach, functioning as 'offices' as well as providing air-conditioning and toilet facilities for family members. Most of them spent the better part of the day there, looking at the remnants of the hotels destroyed in the tsunami, creating makeshift memorials in the sand or having a quiet moment with a minister. The next day, several family members went back, while others had the mini-van take them to the local hospital where surviving family members had been treated.

The official and formal memorial service took place in the afternoon of May 6 on the beach of the Katathani Hotel. The program included poems and speeches, collective songs and music. Before announcing one minute of silence, the Minister of Health, Mr Ansgar Gabrielsen, read out the names of all the 84 Norwegians who had died in the tsunami. For each name, a rose was placed in an ornament formed as a heart, so that when all the names had been mentioned, the flowers represented a red heart on the beach.

A heart was also a central part of the memorial speech given by Hilde Sirnes, a minister from the Norwegian Church Abroad who played an important role in comforting survivors and next-of-kin in Thailand in the days after the tsunami. In beautiful afternoon sunshine on the beach of Katathani, Sirnes described losing a loved one as like having a sharp triangle inserted into your body. 'It turns and tumbles, and its edges pierce sores in your heart. The pain is so enormous you think you will not survive,' Sirnes said. She also gave the audience a feeling of hope, as she continued: 'After days, months or years, the edges soften, and the "triangle of sorrow" starts resembling a pearl. A pearl that will forever stay in your heart. Like a treasure.'

Communication Challenges

The trips to Thailand took place in 2005, before the advent of social media. We therefore had to use 'traditional' means of communicating with participants, including printed fact sheets and instructions handed out on board the plane to Phuket, a manned information booth at the hotel and messages on bulletin boards.

Before family members arrived, I suggested to colleagues in the task force that we produce a printed newsletter that could be handed out each morning. The newsletter would include updated facts about the day's agenda, weather forecast and suggested activities and could be copied in our makeshift headquarters at the Star

FIGURE 2.9 A heart filled with red roses was used to symbolize the casualties of the tsunami on the beach in Thailand.
Source: Kjell Brataas.

Tours office in Phuket. Everyone thought it was a great idea, and not surprisingly, I got the job of writing and producing it. I therefore spent each evening coming up with appropriate text, designing the newsletter in Microsoft Publisher and then waiting for a slow copy machine to churn out 300 copies of the two-page newsletter.

Another idea I had was more rewarding on a personal level. I had spent an afternoon in April at a local tailor near the Karon hotel, and when I picked up my new suits and shirts, the young man who had made my nice clothes mentioned that his brother owned a small bar on the other side of the building. We went to see him together, and we agreed that we would use his facilities for a party for all staff on the very last evening of the last trip in October. I told him we needed his entire bar for ourselves, and that guests could order anything they liked throughout the evening. Not surprisingly, the night at the bar was a great success, and at the end I happily signed a bill that would have been much bigger almost anywhere else in the world.

Fortunately, participants on the two trips were also satisfied. They answered a quick survey on the flights back to Oslo, and the results showed they were overwhelmingly positive about their experience and the organization of the trips. Not surprisingly, maybe, as it was an 'all expenses paid' trip that included flights, connecting flights in Norway, airport transportation, local transportation and all meals for a week.

Communication Challenges in Many Countries

It was not only Norwegian authorities that struggled with handling the tsunami. Several countries in Europe had similar experiences, and quite a few of them were detailed in official reports that came out later. A document by the Accident Investigation Board of Finland stated that Public communications by the authorities were less than successful. The Information Unit at the Ministry for Foreign Affairs restricted itself to official bulletins; these conflicted with the information available to the general public and published by the media, and the latter were subsequently proven correct.

(Accident Investigation Board Finland, 2005)

In the U.K., much of the focus of an official review was on the communication aspect of the crisis:

The experiences reported by survivors in affected areas and by relatives in the UK show that the most consistently difficult and frustrating aspect of their interactions with the UK agencies was the communication of information; mostly (although not exclusively) on missing or located persons.

(National Audit Office, 2006)

BIBLIOGRAPHY

Accident Investigation Board Finland. (2005). *The natural disaster in Asia on 26 December 2004*. Retrieved from www.coe.int/t/dg4/majorhazards/ressources/virtuallibrary/materials/finland/dr3cvra.pdf

National Audit Office. (2006, November 1). *Review of the experiences of United Kingdom nationals affected by the Indian Ocean tsunami*. Retrieved from www.nao.org.uk/wp-content/uploads/2006/11/Review_Tsunami_Experiences.pdf

Queensland Police Service. (2011). Disaster Management and Social Media – a case study. Retrieved from www.police.qld.gov.au/corporatedocs/reportsPublications/other/Documents/QPSSocialMediaCaseStudy.pdf

Rieti, J. (2013, June 25). *CBC News Canada*. Retrieved from www.cbc.ca/news/canada/alberta-floods-keeping-up-with-calgary-mayor-nenshi-1.1303739

Tasmanian Government. (2013). *2013 Tasmanian Bushfire Inquiry Report*. Retrieved from www.dpac.tas.gov.au/__data/assets/pdf_file/0015/208131/1.Tasmanian_Bushfires_Inquiry_Report.pdf

'The National'. (2013, June 21). Mayor Nenshi on Floods [Video file]. Retrieved from www.youtube.com/watch?v=ohVXYmlkLFc

Vroegop, R. (2014). *Forewarned and forearmed: The Calgary Emergency Management Agency and the 2013 flood*. Ottawa: The Conference Board of Canada

CHAPTER 3

Terror

INTRODUCTION

Sadly, terror attacks are becoming more and more common, in the U.S. as well as in the rest of the world. And with social media bringing images directly from the scene to the living rooms and smartphones of the general public, people around the world learn about such atrocities simultaneously and sometimes in real time.

This chapter focuses on terror attacks that took place in Boston, Algeria, Orlando, Oslo and Utoya. Their means of spreading death and bereavement differ greatly at each location, but from a crisis management standpoint they provide numerous examples of the many challenges a terrorist attack creates for victim support, media handling, international cooperation and crisis communication.

BOSTON MARATHON BOMBINGS

Case: Two bombs exploded near the finish line of the Boston Marathon at Boylston Street. Social media were used extensively in the aftermath, spreading rumors and false information as well as corrections and important messages for the safety of the public and police.

When: April 15, 2013.

The Boston Marathon is an iconic race. It is one of the oldest in the U.S., and as it is always held in April, it is a symbol of the City of Boston awakening from winter. The marathon is a celebration of spring, renewal and, of course, energy and invigoration through athletics. Runners are exhausted and enduring some of the most challenging and physical activity of their lives, but at the same time the city has an atmosphere of lightness and positive vibrations.

The race always draws thousands of spectators. Many of them were gathered around the finish line on Boylston Street, enjoying the nice weather and cheering the runners. The elite runners started in the morning, so in the afternoon it was mostly casual runners and those participating for charities who were still on the course, a vast mix of old and young, fit and slow.

The first bomb detonated at 2:49 pm, the second only 13 seconds later. At first there was panic and confusion, but medical volunteers and bystanders quickly responded to the many wounded. At the same time, there was speculation about more bombs, and then came questions about how to inform the runners and the public.

Social Media Ready

In 2012, the City of Boston had appointed Lindsay Crudele as its very first social media director. She initiated an assorted social media presence for the city, which included accounts on Twitter, Facebook, Tumblr, Pinterest, Instagram and SoundCloud. Crudele eagerly trained office personnel in handling social media, which resulted in the city having more than 50 different Facebook accounts and 50 different Twitter accounts for departments and interests ranging from archives to transportation.

Shortly after the explosions, Crudele and her team flipped into an emergency mode, which included a multi-channel approach combined with a common voice that could speak on several social media accounts. Enterprise-style management and account consolidation, put into place in advance, helped centralize the social media accounts, and another quick decision was to remove all scheduled information, so that premade messages about totally different subjects would not appear when much more urgent posts needed to be the focus.

As is typical for breaking news, the first message on social media about the explosion came on Twitter, when only about two minutes after the blasts runner Josh Cox published a picture of the finish line – with smoke from the explosion in the background – saying: ‘Do not go near the finish line at the #BostonMarathon – 2 explosions in buildings.’ It was retweeted more than 300 times.

Official information came quickly, too. Within an hour of the explosion, the Boston Police Department used Twitter to inform the public: ‘Boston Police confirming explosion at marathon finish line with injuries.’ Thirteen minutes later, they tweeted: ‘Updates to follow. Please clear area around marathon finish line #tweetfromthebeat via @CherylFiandaca.’

The Twitter handle @CherylFiandaca showed up in a number of messages from the Boston Police. As bureau chief of public information for the police department, Cheryl Fiandaca had been instructed by Police Commissioner Ed Davis to increase the department’s social media presence. Davis wanted communication with the public through social media, and it was Fiandaca and her team of five who were responsible for social media messages from the police during the Boston Marathon. They didn’t waste any time after the explosion, and Fiandaca tweeted the first message after the explosion from a shopping mall (Bindley, 2013).

There was also an official Twitter account for the Boston Marathon. About two hours after the detonations, they published a tweet about where runners could meet their families. To find the locations of the runners, the public could also use a website with ‘checkpoint tracking,’ where information was stored about which runners had been stopped from continuing to race after the explosions.

FIGURE 3.1 Twitter was used for a variety of purposes, including showing a personal side of the Boston Police Department. *Source*: Twitter.

An Official and Visible Voice

As it happened, the mayor of Boston, Thomas M. Menino, was in hospital recovering from surgery on April 15. It was important to show Bostonians that someone was in charge, and that the local government was still there and in action. Menino checked himself out of the hospital to show control through presence, and he also took part in several press conferences during the next days.

Menino could very well have been a victim himself, as he described in an article a year later:

> For 19 years I sat in the front row of the bleachers, right across from where it happened. I immediately thought about that. I could have been part of the injured. I used to sit there with all my grandkids. I didn't panic. I kept on saying: We gotta make sure the public knows what really happened. We're in control. We're not going to let the terrorists take over our city.
>
> (Menino, 2014)

The City of Boston and the Boston Police collaborated on press briefings and press releases. The communications team often started their day with a meeting at 6 am, then held a press briefing – followed by more briefings during the day if there was developing news or vital information to give out. In the days after the bombing, the social media team of the City of Boston continued and expanded their use of 'shareable social graphics' – short messages in the form of a picture with text or a graphical element. They customized content for each social platform and used a variety of tools to visualize trends and data. Hootsuite Enterprise became their most important digital dashboard, but Dataminr and GroundSignal were also of great help in trying to get an overview of what was happening.

Misinformation and Rumors

The Boston Marathon bombings became an example of how incorrect information can spread rapidly – on social as well as traditional media. CNN, AP and the *Boston Globe* all reported stories about an arrest that had not taken place, and the *New York*

Post wrote that 12 had died in the attack (even citing law enforcement as its source).

From a few minutes after the explosion, social media, especially Twitter and Reddit, were filled with rumors, hypotheses and a 'hunt' for the perpetrators. Bad tips were circulated, the library was mentioned as the original target, and the names of several innocent people were mentioned as possible suspects. Undoubtedly, this information did not help law enforcement and at the same time spread unnecessary fright and paranoia.

Reddit's digital 'voting system' was initiated to try to find the bombers. On the website, the public was invited to scrutinize hundreds of pictures in order to see if they could identify suspicious activity or the actual bombers. Although hundreds of people took part, it did not succeed, and several commentators saw it as a failure of modern crowdsourcing. (The main reason for the unsuccessful attempt was that pictures of the two brothers who organized the bombings were never made public; Surowiecki, 2013.)

The Reddit attempt to find the criminals through crowdsourcing unleashed a vigorous debate and criticism regarding the use of social media. It also exemplified that people are much more willing to share rumors or spread incorrect information when they are not in a face-to-face situation, and that comments under the pictures on Reddit resulted in many people following a trend instead of thinking independently. In an article in *LA Times*, a media studies professor at the University of Virginia, Siva Vaidhyanathan, commented: 'This is one of the most alarming social media events of our time. We are really good at uploading images and unleashing amateurs, but we're not good with the social norms that would protect the innocent' (Chang, 2013).

Police Radio Open to the Public

As the hunt for the suspects continued, individuals all over Boston – and the world – could follow police actions through streams of online scanners. Many tweeted what they heard, and tactical information about police actions was therefore spread on social media. Such messages obviously put officers at risk, and on April 19 the Boston Police Department issued a warning on social media: '#MediaAlert: WARNING. Do Not Compromise Officer Safety by Broadcasting Tactical Positions of Homes Being Searched.'

The frenzy of social media messages with detailed information about police whereabouts, rumors and comments received lots of attention, and traditional media struggled to keep up with the news. At one point an individual commented on Twitter:

FIGURE 3.2 Social media were the preferred news channel for many people following the manhunt after the Boston Marathon bombings.
Source: Twitter.

Taking Care of Survivors

The City of Boston organized a family center at the Seaport Hotel. There, next-of-kin of those who had died and those who were hurt could gather and get updates. The city also set up a closed Facebook group for survivors, which proved to be an effective way of communicating with the people most directly affected by the explosions.

Google also launched its system for finding people after a tragic event: Google Person Finder. Through this free and easily accessible tool, the public could use Google's search muscles to look for their loved ones – or report about their own safety (similar to today's Safety Check from Facebook).

The FBI also sprang into action to take care of victims, and their Office of Victim Assistance set up a web page that encouraged those who were injured or who had witnessed the bombing to register. The site also provides links to a variety of resources, including hints on coping with trauma and guidelines for helping youth after the bombings.

A Visit from the President

With only 19 hours' notice, the City of Boston also had to organize a visit from President Obama, resulting in hectic activity and several meetings with the Secret Service. When they arrived, the President and his entourage first took part in a memorial service at Cathedral of the Holy Cross. The service also included comments by Mayor Thomas Merino and Massachusetts Governor Deval Patrick. The president then met privately with patients, families and hospital staff at the Massachusetts General Hospital, while Michelle Obama visited Boston Children's Hospital and Brigham and Women's Hospital (Eversley, 2013).

Funding Initiative

To give Bostonians – and people all over the world – something positive to focus on, the City of Boston quickly organized a fund for receiving and organizing donations: One Fund Boston. Mayor Menino and former Governor Deval Patrick were personally involved in its inception, and less than 24 hours after the attack the fund was initiated – with papers filed and new bank accounts linked to PayPal.

Dot Joyce was the former chief communications officer and press secretary of the City of Boston. 'We started the One Fund Boston with messages of hope and resilience. We saw it as an outlet for anger and frustrations, and we also wanted to focus on togetherness and something positive children could do in the aftermath of the bombings,' Joyce said about the initiative.

The City of Boston partnered with Twitter, which provided a pro bono 'Promoted Trend' campaign for digitally launching the One Fund. It immediately became a success, and tweets about the fund spread rapidly. The mayor also used his personal Twitter account to promote the fund and raise awareness, tweeting: 'To assist victims of the #bostonmarathon tragedy, please visit onefundboston.com #oneboston.' The message was retweeted more than 700 times.

The One Fund has become a 'best practice' example of how to collect funds after a tragic event. There were many reasons for its success, but some vital clues were the speed of launch, messages on a variety of social media channels and a large initial donation. A sponsor of the Boston Marathon since 1985, the investment company John Hancock donated USD 1 million to get the One Fund started, thereby setting an example for other sponsors and companies (Weiss, 2016).

By the first week, the fund had USD 10 million in its account. The One Fund set up a system which asked people to self-identify as victims of the attacks, and with money collected so quickly, the fund was able to start issuing checks to the injured within a short time. About USD 60 million was initially disbursed to survivors and families of the victims. There were no spending criteria. When USD 20 million more came in, the money was given to survivors and a few organizations that could provide care for the victims.

In order to ensure a speedy distribution of funds, the One Fund management team used a category model for who would receive money from the fund, with a descending amount of support:

- Families of the four people who died and the two people who suffered double amputations
- Persons who suffered single amputations
- All others who had been hospitalized for at least one night, with compensation determined by length of required hospitalization
- Persons who suffered physical injury and received emergency room treatment but were not admitted for hospitalization

As planned from the start, the One Fund shut down and stopped collecting donations 20 months after the terror attack. By then, it had received contributions from more than 200,000 individuals, groups and businesses.

The Boston University's Initiative on Cities published a report in 2014 about lessons learned from the One Fund initiative so that 'communities affected by mass trauma who are confronted with a similar desire and need to coalesce empathy into a useful form' can learn from Boston (Yesnowitz, 2014). The report lists five key ingredients for the success of One Fund Boston:

1. Strong political leadership
2. Establishment of a single disbursal engine
3. Speed of execution

4. Preexisting relationships between public and private sector leaders
5. Commitment to transparency and community involvement

One Boston

Shortly after the attack, the City of Boston started receiving condolences and pictures from all over the world. The social media team therefore developed a presence on the photo sharing site Tumblr, and used a service called IFTTT (IF This Then That) for automatic transfer of pictures to the blog. 'One Boston' was a common theme, as was 'Dear Boston.'

Since 2015, April 15 has been dedicated as a day that encourages public service, charity and community resilience. The hashtag #OneBostonDay has become a popular tool for sharing how the day is being observed, and the City of Boston has even produced a number of shareable social graphics that can be downloaded for free. The One Boston Day website describes the day like this: 'On this day, we remember and reflect. We greet our neighbors. We lend a hand. We reach out, give back, and go above and beyond. We epitomize the spirit of the city we love.'

IN AMENAS ATTACK ON GAS FACILITY

Case: Terror attack on gas facility at In Amenas, Algeria. Forty people from 10 countries lost their lives during the four days' siege. Among them were five employees of Norway's largest company, Statoil.

When: The attack started at 5:30 on January 16, 2013.

The attack on the Statoil facilities in Algeria was the most serious international crisis the company had ever faced. There were many challenges regarding communication, but Statoil handled most of them well and was later applauded for the way it informed and supported the families of those employees affected by the crisis.

With more than 800 people on site, there were many uncertainties from a Statoil crisis management standpoint during the first few hours of the attack. In Amenas is in the desert and far away from any cities, and its remote location meant that few other sources – such as local media – were reporting from the incident. A positive side to this was that there were fewer rumors and less potentially false news developing out of a grave situation. Nevertheless, because of the scope of the terror attack, and also the fact that there were workers at In Amenas from 29 different countries, the attack immediately erupted into a global news crisis.

The leadership of Statoil, represented by the CEO, Mr Helge Lund, proved to be a vital part in the handling of the event. On the morning of January 16, Mr Lund arrived on a business trip in Tokyo, where he was given a quick briefing on the situation in Algeria. He decided to return immediately to Norway and the Statoil headquarters in Stavanger, where he joined the emergency response team at 00:20 on January 17.

Crisis managers with Statoil had learned several lessons from the *Deepwater Horizon* disaster in the Gulf of Mexico, and on January 16 they immediately planned for the worst. Internal authority for handling the situation was quickly handed over to the Statoil hostage incident response team, and they then made contact with outside sources, partners and government agencies such as the police and the Ministry of Foreign Affairs.

Three days after the attacks had started, Algerian military took control of the production area at In Amenas. The next day, Statoil received confirmation that 12 of the 17 employees missing were safe. After family members had been informed, the names of the five people missing were made public on January 20.

Family Support

Statoil is known for supporting families affected by a crisis, and about four hours after getting the first message about 'something' happening at In Amenas, a telephone support system was in place. The support staff soon realized the terror attack could last a long time and possibly mean casualties, and a next-of-kin center at a hotel in Bergen was ready within one hour after being contacted. The first family members arrived in the afternoon of January 16. The next day, Statoil took over the whole hotel, and it remained a family center for eight days.

Being the largest employer in Norway, and with a major focus on health, safety and environment and good values, Statoil had a robust system in place for taking care of next-of-kin. A hundred employees had been trained as 'volunteers' for family support, and they came from all kinds of backgrounds and fields of work. Many of them worked in HR, but some volunteers were also engineers and accountants. The management of Statoil sees this kind of mix as highly valuable, as the volunteers then have a varied background and a diverse knowledge of Statoil personnel and operations.

At the Scandic hotel in Bergen, a predefined plan for use of the facilities was set in action. Access control, registration and group rooms were organized, and during the week more than 100 Statoil employees were involved with family support at the hotel.

The main reason for gathering family members at the hotel was to be able to give them firsthand information and news at one place simultaneously. It was important for Statoil that this was done in a professional and correct way, and they therefore set up a rotation scheme for top managers, who alternated the difficult task of giving bad news – or sometimes no news at all – to family members of those missing or dead. Mr Helge Lund took his share of giving updates at the next-of-kin center. 'Families, friends and colleagues are waiting for news about their loved ones. We are doing everything we can to assist and help all of them,' Lund said during a press conference about the situation.

Statoil also organized several ceremonies to show respect for those missing or known dead. The company held a mourning ceremony at Haakon's Hall in Bergen on February 4, and in its invitation Statoil stated that 'It is intended for the next of kin, those directly affected, representatives of management, support staffs and other Statoil employees involved.' The same day, a one-minute silence was planned for Statoil locations all over the world. Condolence books were placed in central Statoil locations, and thinking of its very international staff, a digital book of condolences was also organized.

Communicating with Media and Staff

Statoil's management team, as well as government officials, made it an important point to always give information to family members before giving it to the press. At the same time, there was enormous interest from journalists, and Statoil therefore arranged several press conferences where the CEO and other key people from management were available for interviews. The Statoil communications team also set up a dedicated web page, which included the latest facts and press releases, as well as pictures and maps available for download.

Statoil issued their first press release in the early afternoon of January 16. It simply stated that their plant had been attacked and that they were gathering information. In total, during and after the event, Statoil published 26 press releases and held nine press conferences.

It is very seldom that a crisis takes place in a location where no outside witnesses or the public report about what has happened. This was the case at the plant at In Amenas, and the communications team of Statoil therefore became one of the only sources of information for the media.

Contrary to most of the other case studies presented in this book, social media did not play a very important role in handling the In Amenas crisis. The communications team in Norway used various tools for surveying social media in several countries, but social media were not used extensively to get messages out. Statoil's main objective was to communicate directly with those affected and their families, and they gave updates to the media and the public several times a day.

Providing information about the crisis to Statoil employees in a total of 35 countries took various forms. An effective way of doing this was a town hall meeting that took place in Stavanger. Statoil's CEO Helge Lund and Norway's prime minister Jens Stoltenberg attended the meeting. It lasted for 30 minutes and was broadcast live to Statoil employees all over the world.

An Official Report

About a month after the terror attack, Statoil's board of directors commissioned an investigation into the attack. The investigation team submitted their report on September 9, 2013. It stated that extensive resources were made available, the teams involved had freedom to act, and survivors and next-of-kin gave positive

FIGURE 3.3 Statoil and the Office of the Prime Minister worked jointly on the In Amenas crisis. Prime Minister Jens Stoltenberg (left) and Statoil's CEO Helge Lund both spoke at a staff meeting. *Source*: Harald Pettersen, Statoil.

feedback regarding the support they had received. 'Statoil's contribution to the overall emergency response was effective and professional [...], with clear and honest communications,' the report concluded (Statoil.com, 2013).

Facts about the Terror at In Amenas

- The In Amenas facility was operated as a joint venture between the national oil company Sonatrach, BP and Statoil.
- In Statoil alone, more than 350 people were mobilized as part of the company's emergency response effort.
- Statoil published a report on the attack and how the company handled it in September 2013.
- The attacks generated 9,000 news articles in Norwegian media.

TERROR IN THE CAPITAL OF NORWAY AND ON THE ISLAND OF UTOYA

Case: A terrorist attack on the Government Complex in Oslo and a youth camp for the Labor Party on the small island of Utoya. In Oslo, eight people were killed when a bomb exploded outside the Prime Minister's Office; at Utoya, the terrorist shot and killed 69 people. Internal as well as external communication became difficult and partly broke down in the early phases, as the bomb hindered normal communication tools and made the situation highly uncertain and unmanageable.
When: July 22, 2011.

It happened on a Friday. July 22, 2011. I was happily driving home from vacation in Sweden with my wife and three kids when all of a sudden I started getting several text messages from colleagues in Oslo. 'Are you OK?' they asked, and I at first thought they were requiring about my vacation. Then a text message from my boss said something about 'I hope you are all fine, please report back as soon as possible.' I stopped the car, sent her a short message, and then we turned on the news on the radio. We were immediately stunned. The reporter informed of an explosion in Oslo. It was too early to tell whether there were casualties or what was the cause. We drove in silence, listening to the continued coverage on the radio.

Once safely back home in our house, we immediately turned on the television to watch live coverage from the Government Complex in Oslo (where my office was). The pictures were unbelievable: chaos, debris and glass fragments, wounded people and dust everywhere – and the building where I worked was on fire.

After the tsunami in 1984 (see Chapter 2), the Norwegian government had organized a crisis communication support group consisting of communications specialists from several ministries. I was one of their 'volunteers,' and I therefore knew that I would probably be requested to come to work fairly quickly. Rightly so; in the early part of the evening I got a phone call urging me to come to Oslo as soon as possible to help with communication efforts after the attack. There were two obstacles, however. First, the crisis team was supposed to meet at the Crisis Support Unit, but their facilities were in the basement of the building that had been bombed in Oslo and therefore unreachable. Instead, I was instructed to report to work at the Ministry of Defense, as their offices were further away from the Government Complex and therefore easier to reach – and protect. The second challenge came in the form of a news flash popping up on TV as I was about to leave in a taxi, heading for Oslo – there was something about several people being shot at Utoya. This puzzled me, and I started to wonder whether we also had to deal with a situation there that needed extra communications personnel. An already bad feeling further developed and was unmistakably reinforced when minutes later – still in the taxi towards Oslo – we encountered several ambulances driving fast in the opposite direction. I later learned that they were on their way to Utoya.

The Office of the Prime Minister

In hindsight, it is easy to see that security around the Government Complex was too relaxed. Anyone was, for example, able to drive a car near the entrance and park it there: exactly what Anders Behring Breivik did. He parked his Volkswagen Crafter containing his home-made bomb (consisting of 950 kilos of fertilizer, diesel and aluminum), left it there and walked out into the streets – wearing a fake police uniform. Security personnel quickly started looking into the situation, trying first to find the owner of the car by checking its license plate via video surveillance. However, no immediate attempt was made to remove the car, and when it exploded it instantly killed eight people on the ground and wounded more than 200.

Fortunately, Prime Minister Jens Stoltenberg was not in his office on the top floor of the building when the bomb went off at 3:25.19 in the afternoon. The blast rocked the 17 stories of the structure, and staff immediately started evacuating through the stairs. One of the communications people in the Office of the Prime Minister immediately called Jens Stoltenberg and told him that there had been an explosion and that they were running out of the building. They agreed to meet at the Residence of the Prime Minister, where Stoltenberg had spent the afternoon preparing a speech he was to give at Utoya the next day. Within a fairly short time, several staff reported to work at the residence, including communications staff who had had the great foresight to bring

FIGURE 3.4 Jens Stoltenberg, Merete Guin and Arvid Samland worked in makeshift offices at the Residence of the Prime Minister on the evening of July 22, 2011.
Source: Office of the Prime Minister.

their personal printers. This proved valuable, as the residence was not set up to handle lots of office workers. The training room was used for sorting papers, and during the afternoon and evening several 'official' rooms were turned into makeshift offices.

Handling the Media

Immediately, the media pressure was enormous. In the building where I had my office (R4), a colleague got a call from a reporter while evacuating through the emergency exits, and journalists from all over Norway – and soon the world – were trying to find out what had happened to the prime minister. An international news agency got hold of the phone number to one of Stoltenberg's advisors, called him up and said: 'You are live in 20 seconds.' For a while the press line was so busy that it was impossible to use it to make outside calls.

The Prime Minister's Office soon released a statement saying that Stoltenberg was safe, but they made sure not to mention anything about where he was located. There were several reasons for not disclosing his whereabouts, the most important being that at this time no one could know whether there were more attacks coming. At the same time, Stoltenberg wanted to address the media and get his messages across so that he could reassure the public and not risk others dominating the media with rumors or false information. They therefore organized telephone interviews with Norway's two major TV stations (NRK and TV2) in which Stoltenberg talked about what actions were being taken, that a crisis team had been established and that everything was being done to save lives.

By this time, Stoltenberg knew that something was going on at Utoya. Representing the Labor Party himself, he personally knew many of the young people attending the youth camp, and he had received several messages about someone shooting on the island. His team discussed what to say to reporters if they should ask about Utoya, and they agreed he would be honest and inform them of what he knew if they asked. At the end of the interview with one of the news channels, the reporter said that he had heard that 'something' was happening at Utoya, and he asked what Stoltenberg knew about the shooting. Stoltenberg said that he had received reports about someone shooting young people on the island. This turned out to be vital and possibly life-saving information, as neighbors of the island who watched the broadcast then realized people were in danger and subsequently set out in small boats to try to save people attempting to swim away from Utoya.

Norway's Response to Violence: More Democracy, More Openness …

Hans Kristian Amundsen was one of Stoltenberg's closest advisors and responsible for most of his speeches. When the attack occurred, he was vacationing in the northern part of Norway, and he immediately booked a ticket on the first flight back to Oslo. He therefore had one and a half hours of undisturbed time on the plane, which he used to draft the first official message from the government about the terror attacks.

Stoltenberg and Amundsen agreed that this was not the time to ask for revenge. Instead, their focus was to show togetherness and that even a small country like Norway could handle a terrible situation in a good way. In his press conference at 10.30 in the evening of July 22, Stoltenberg therefore talked about the fact that Norway had experienced its most brutal attack on democracy since the Second World War, and that 'Norway will not be intimidated by terror attacks.' He condemned the terrorist's brutal assault and said that the country would 'stand firm in defending their values.' Stoltenberg ended the short press conference by stating that 'Norway's response to violence is more democracy, more openness and greater political participation.'

Communicating from Utoya

In the afternoon of July 22, a total of 564 people were gathered on the island of Utoya. As part of a youth camp organized by the Labor Party, they were in good spirits after a visit the same day by former prime minister of Norway, Gro Harlem Brundtland. She left the island at 3 pm, and half an hour later news spread about the explosion in Oslo. The camp therefore organized a brief gathering at 4 pm, during which several participants said they were happy to be so far away from the terrible events in Oslo.

What they did not know, of course, was that Anders Behring Breivik had not finished his plot. Still dressed in a police uniform, he approached the dock that connects Utoya to the mainland and asked a guard there to summon the ferry to take him over to the island. This was organized, and Breivik got on board for the ten-minute ride to Utoya. He had with him a Glock brand semiautomatic 9 mm pistol, a Ruger brand Mini 14 semiautomatic rifle cal. 223, a heavy box, and a sharp and visible bayonet. At 5:17 he climbed ashore, and four minutes later he killed two people who were with him on the dock. This was the start of a terrible killing spree; during the next one hour and 20 minutes, Breivik shot and killed 69 people. He was arrested by police at 6:34 pm.

FIGURE 3.5 A dramatic message on Twitter: 'Someone is shooting at Utoya.'
Source: Twitter.

One of the first messages about what was happening at Utoya came on Twitter, when at 5:25 Kjetil Vevle – a 23-year-old who was one of the first to see what was happening – reported: 'Shooting at Utoya. Update the police!' Several similar postings followed, and at the same time numerous police operations centers were inundated with frantic calls from the people on the island.

For the crisis communications team gathered at the Ministry of Defense, the frequent messages on Twitter made it hard to keep on top of the situation and describe to crisis managers what was happening. Traditionally, such reports are based on what police and other agencies provide combined with media surveillance, but live tweets and text messages from Utoya meant that many of the politicians and key people handling the crisis got their updates directly from those who were at risk.

News about the shooting came sporadically and then spread quickly. Norwegian reporters knew several of the people on the island and called them directly, but international journalists had to call a dedicated phone number set up specifically for the media. I was among those who answered these calls, but there was not much information I could reveal, at least not in the early hours of this dreadful happening.

> **Descriptions from the Judgment**
>
> "Heartbreaking scenes unfolded as people hid, ran or swam for their lives while at the same time trying to help and comfort each other. In some places, the living and the dead lay side by side. Some were paralyzed by fear as they were being shot at, some pretended to be dead, while others begged for their lives. Many hyperventilated. Telephone calls were made and text messages were exchanged with family and friends, partly to ask for help, partly to calm them down, and partly also to bid farewell.
>
> Oslo District Court (Oslo tingrett) – Judgment.

For a long time that first evening, the police reported that about ten people had been killed on the island. However, politicians and crisis managers had seen indications or even text messages from witnesses on the island that had shown much higher numbers, and at 3:50 the next morning the police told the media that 'at least 80' had died on the island. Needless to say, Norwegians woke up that morning to once again shocking news, and those in Oslo saw a very different city, with police and the military guarding key buildings and offices in the capital.

Taking Care of Survivors and Next-of-Kin

The all-important challenge was then to handle those who had survived the attack and to get vital information to next-of-kin. This was not an easy task, as the sheer number of people involved meant that patients were transported from Utoya by private boats, private cars, helicopters, ambulances and buses. A total of 12 helicopters, 42 ambulances and three buses took part in the evacuation of those who had been on the island.

While survivors and deceased were taken to several hospitals, family members started gathering at the Sundvollen Hotel, situated fairly close to Utoya. The hotel did its utmost to provide care and necessary rooms, and during the evening all the regular guests were moved to other hotels in the area. Parents, sisters, brothers and other family members came from all over Norway, and many of them wanted to take part in the search and rescue efforts and help locate their loved ones.

Although this happened in the middle of summer, with lots of Norwegians vacationing abroad, more than 250 medical personnel, psychologists and nurses reported to Sundvollen Hotel in the evening of July 22. Organizing them into groups and noting their specialties took some time, but their assistance proved invaluable as more and more parents and other family members showed up at the hotel. There was chaos and loud voices at first, but as the number of dead became apparent, the hotel became very quiet and mournful.

On Saturday July 23, the prime minister held a new press conference at the residence. He called the attack a 'national tragedy.' Then he traveled to Sundvollen Hotel to console and meet with family members. King Harald and Queen Sonja of Norway also visited the hotel that day. Such VIP visits are usually highly appreciated by family members and should therefore be encouraged, but they put a special demand on crisis communication, security and logistics, and it is crucial to also plan for these situations after a tragic event.

Internal Communication

It was not only the media and the public who needed information about what was happening. The Government Complex in Oslo houses 11 different ministries, and many employees were directly affected when the bomb exploded. Office workers who had been at the site needed to be taken care of, and at the same time many on vacation wanted to know what was happening and whether their colleagues were safe.

Communicating internally proved difficult at first. The bomb had destroyed several IT systems, which meant that for a while intranets and email did not work. Communications staff at the ministries had to find alternative ways of getting information to their employees. Regular phone calls and text messages were used, and by the evening of Friday 22, several ministries had created closed Facebook pages. This proved very effective, as employees could comment and ask questions, and information could be distributed freely and easily to only those who needed it.

It soon became apparent that employees wanted to meet each other in person. Many ministries therefore organized open meetings for their staff as early as Sunday July 24. I helped set up a meeting for the Ministry of Trade and Industry at a hotel in Oslo, and we were quite surprised to see how many showed up that afternoon. It was a relief to learn that no one at our ministry had been killed, but of course, at the same time there was sadness and grief about what had happened. Many personally knew some of those who had not survived the explosion. The main reason for organizing the meeting was simply to let staff see each other and be able to talk and cry together, but at the same time many practical questions came up: Some wanted to know how they could help; others were wondering where to show up for work the next day. We also learned that many employees had kept a variety of personal effects in their offices, and there were questions about how they would obtain their passports, keys to apartments and even national costumes that had been stored in the office closet. And, of course, there were questions about the cars they had driven to work on July 22, as the parking garage was directly below where the bomb went off.

According to the official report published by the Norwegian government in August 2012, only four of the 17 ministries had a predefined alternate location. There was therefore hectic activity to find new office space for the 2,000 or so employees who could not report to work at the Government Complex the following Monday. This was also a communications challenge, as staff needed to know where to come to work and what to bring.

Employees of the ministries showed a variety of reactions. For many of them, their office had been blown to pieces, they had been forced to evacuate in dramatic circumstances, and they had seen wounded or dead colleagues. Heads of the ministries therefore offered referrals to health workers and psychologists, and everyone was told that they did not have to show up for work the next day if they did not feel up to it.

Memorials and a Sea of Roses

A common reaction after a tragedy is the need to come together and show support. An individual suggested on Facebook that there should be a ceremony in the city of Oslo on Monday July 25, where people could bring torches and roses and walk silently through the downtown area. Twenty-four hours after the event had been launched on Facebook, some 30,000 individuals said they would attend.

Officials in Oslo soon realized that a march through the city would not be feasible, as many areas were still considered a crime scene, and the safety of such a big group could not be guaranteed. Instead, a 'rose march vigil' was organized outside the town hall. Although it was a 'spur of the moment' event, it is estimated that 150,000 people showed up that evening for a silent and somber ceremony. The prime minister and members of the royal family attended, and Crown Prince Haakon told the crowd: 'Tonight the streets are filled with love.' Prime Minister Stoltenberg also addressed the crowd. Thousands of people carried red roses as a sign of remembrance. The English newspaper *Express* commented: 'In a country of 4.8 million, where a single murder makes front-page news, the solidarity rally was probably the biggest since World War Two' (Fox, 2011).

When the ceremony was over, participants were urged to 'decorate the city' with the flowers they had brought. They did. For the next weeks, inhabitants and visitors to Oslo could see roses everywhere, and they instantly became a symbol of what had happened, of remembrance and of democracy.

Without this having been planned, the area outside of the Oslo Cathedral became the main place for laying down flowers, teddy bears and letters. It started slowly, but soon the ground was covered in flowers and other items people put down. It hindered traffic, but the area became a kind of attraction in itself, and more and more people wanted to participate and see it. The Oslo Municipality got phone calls from Norwegians abroad who urged officials not to remove the flowers before they had been able to put theirs down, and the 'memorial' was allowed to remain through July.

In several presentations, the head of communications for Oslo Municipality, Erik Hansen, has said that 'We did not have a communications plan for removing a sea of flowers.' But it had to

FIGURE 3.6 'A sea of roses' in downtown Oslo.
Source: Kjell Brataas.

be done, and at nighttime at the beginning of August gardening crews from the Agency for Waste Management in the City of Oslo started removing the flowers, letters and candles from the area. They did so respectfully; only small trucks carried the items (a total of 33 loads) to an area where they were sorted and cleaned. The National Archive received 4,000 items and 25,000 documents relating to the terror in Oslo and Utoya, while the collected flowers were composted. The resulting soil would later be used for a memorial site.

Back to Where It Happened (II)

Escape from Utoya Island in the evening of July 22 had been hurried and chaotic. Some of the survivors did not want to ever set foot on the island again, but quite a few expressed a desire to go back and relive the terrible, death-facing hours of the attack. At the same time, many next-of-kin of the 69 who had died also wanted a chance to visit Utoya, to learn about what had happened to their family members who had died on the island.

The Norwegian Directorate for Civil Protection (DSB) was therefore given the task of arranging visits to Utoya on August 19 and 20. The planning phase was short and hectic, and involved getting the island ready for a large number of guests of all ages and physical conditions. To make the pathways safer to walk on, tons of gravel was transported by boat and helicopter, and the Norwegian Civil Defense arranged for large tents, trashcans and portable toilets to be located around the island. Parking facilities on the mainland were almost non-existent, so the organizers decided to transport guests by bus from Oslo. A special vessel was designated for bringing large groups of people across the lake to the island, and other boats and ferries were also made available.

Tore Drtina from DSB was head of the organizing committee. For the last two weeks before the visits, he attended daily meetings in Oslo with representatives from the support group, the Red Cross, the Ministry of Justice and the police. 'The planning phase was intensive and emotional,' Drtina explained. He believes an important factor in the success of the project was that planning was inclusive and that many voices were heard and considered. 'We knew that our guests would be anxious and worried, and we therefore did everything we could to make the visits to the island as smooth and well-organized as possible,' Drtina said. He commends everyone involved, especially the police National Criminal Investigation Service, which produced detailed maps of Utoya (including the location of victims) and took great care in providing adequate signage on the island and personal care of the visitors.

Media handling became an interesting challenge. There was a huge international and national interest in the two days of visits to Utoya, and organizers had to balance the media's rights to cover important events with survivors' and family members' right to privacy. In the planning phase, survivors had made it very clear that they did not want to be surrounded by journalists and photographers on the island, and a 'media camp' was therefore organized on a farm across the lake from Utoya. Large tents were erected in case of rain, and reporters from CNN, BBC and other international and local news organizations were invited to cover the Utoya visit from the camp.

DSB cooperated with the police regarding how visitors would see the island (still considered a crime scene) and remnants of what had happened. Blood had been removed from walls and floors and the houses on the island had been cleaned, but other evidence of what had happened – such as bullet holes – was still present when visitors arrived. That is how they wanted it, and seeing the physical

FIGURE 3.7 From a temporary media camp, reporters could watch across the lake as visitors arrived at Utoya.
Source: Kjell Brataas.

evidence of what had taken place gave an extra dimension to the experience.

Before the guests arrived, DSB conducted a thorough risk analysis. Many factors had to be considered, including the varying age of the visitors, the physical condition of the pathways, and the possibility of boats colliding while transporting visitors or people falling in the water.

August 19 was set aside for family members representing 50 of the young men and women who had died on Utoya. Upon arrival, they were met by representatives from the Red Cross, and each family had a designated policeman who pointed out where their loved one had been found. Many lit candles and left flowers in makeshift shrines around the island.

The next day, survivors of the mass shooting returned to the island for the first time after the attack. As there were 495 of them, each could only bring one family member or friend along for the visit. They started arriving early in the morning, and survivors spent many hours on the island, wandering around and talking to each other.

DSB used the Sundvollen Hotel as their local headquarters for organizing the visits. One meeting room was designated for communications staff, and I was one of the people from the government handling media and surveying social media. We wanted to be aware of possible complaints or questions on Twitter, and I therefore spent most of the day at the hotel following Twitter streams and hashtags such as #Utoya via the social media management tool Hootsuite. There were few complaints, but when someone posted a message about being fearful of meeting journalists on the island, I could send a reply ensuring them that no reporters were allowed on Utoya.

In total, more than 1,600 individuals visited the island on August 19 and 20. All their expenses were paid for by the government, including connecting flights to Oslo, bus and boat transport to Utoya, and meals and refreshments on the island.

FROM NIGHTCLUB TO NIGHTMARE IN ORLANDO

Case: A gunman opened fire at a gay nightclub called Pulse, one of the biggest nightclubs in Orlando. Forty-nine people were killed. For the first time ever, Facebook launched its Safety Check feature for an event in the U.S.
When: June 12, 2016

It was supposed to be a happy, Latin-themed night at Pulse. More than 300 people were inside when, at around 2 am, shooting started. Shortly afterwards, the nightclub posted a warning on Facebook urging 'Everyone get out of pulse and keep running.' The ordeal lasted several hours as the gunman used his AR-15 type assault rifle and a handgun in heavy gun battles with law enforcement. In the end he was shot dead by the police, but with 49 innocent people killed, the Pulse tragedy became the worst mass shooting in the U.S. in the past 25 years (until the shooting in Las Vegas in October 2017).

The Pulse shooting is another example of victims using social media to communicate *while* the ordeal was going on. According to the report 'War and Tweets,' victims provided information that facilitated the rescue and communicated with the police as well as friends and family from inside the nightclub (Burke, Sims, & Sterman, 2016). Approximately 20 to 30 people hid in a bathroom for four hours, and several of them sent videos and pictures through social media. The report states that few videos made it onto news media, as news directors decided not to air graphic material from the victims. However, three video clips originally posted on Snapchat by a victim who did not survive the attack were later uploaded to a Facebook page, garnering more than one million views in 24 hours and 3.7 million views by the end of October. The video also made its way to television news stories.

With several large hurricanes passing through Florida for the last few years, the City of Orlando had learned several lessons in crisis communication. It had implemented mass casualty incident (MCI) training with the Orlando Police Department and the FBI as early as 2005, and in 2015 the city took part in an active shooter exercise on a college campus.

Social Media for Crisis Communication

One part of the crisis plan for Orlando stated that in the case of a large event, 'ESF-14' – a group of communications people from various emergency operations functions – would work together on crisis communication. Shortly after the shooting started, the Orlando Police and the City of Orlando were overwhelmed with hundreds of calls from the media. The communications team were not able to answer them all, and they therefore used Twitter as their main channel for putting information out to journalists and the public. The Orlando Police Department handled their Twitter account, but ESF-14 took over all other accounts of the city and engaged partners to share messages throughout the night.

Doug Richards, digital communications manager with the Office of the Mayor in Orlando, explained that 'the big thing following the incident that morning was that Twitter allowed the Orlando Police Department to provide accurate information. That really was the lone source for accurate information.' Again, social media proved to be a powerful tool, and the successful use of Twitter by the Orlando Police Department showed the importance of having an already very active and credible Twitter presence.

Two elements of the shooting created an extraordinary number of hateful comments on social media: Pulse was a gay arena, and the killer was a Muslim. In the early hours of the night, the communications teams of the city tried to monitor all the comments, but as more information about the shooting came out, that became an impossible task. Fortunately, there were also many positive comments, and questions soon started appearing about how individuals could help or where they could donate money.

Sharing graphics on social media helped in getting important messages across to a vast audience. The City of Orlando brought in their graphics team early, and they were tasked with producing signage, website design, road closure maps and backdrops for media interviews. The graphics team also developed a new logo, and the hashtag #OrlandoUnited was featured prominently in graphics and messages on social media.

Handling the Media

In the hours and days following the shooting, the Orlando police arranged several press conferences. The first took place at 7:15 in the morning, about five hours after the shooting had started. The 'audience' were mainly journalists, but in order to reach more widely, the police broadcast the conferences live through the Periscope app. The initial plan for the press conference was to have the police chief speak first, but before the conference started the order was changed so that Mayor Buddy Dyer took the podium first. He used the opportunity to set the tone and to let people know they were safe. The second press conference was more solemn, as the mayor then had bad news to deliver: 'Today we're dealing with something we never imagined and is unimaginable. Since the last update, we have gotten better access to the building. We cleared the building. And it is with great sadness that I share we have not 20, but 50 casualties' (Burke, Sims, & Sterman, 2016).

The City of Orlando created a special page on their website with content related to Pulse. It included all relevant updates and special sections such as volunteer forms and public records requests. The website was the location for accurate information, and several messages on social media referred to the web pages for further and detailed facts about the shooting. The web also became an important source for journalists.

In most cases it is the police who release names of the dead to the media and the public. The Pulse tragedy was handled differently, as city staff were tasked with publishing names of the deceased on the city's web page after police had notified next-of-kin. The 'War and Tweets' report explains that 'an assistant would eventually take calls throughout the night from an officer who was notifying victims' next of kin. Once the web assistant received that information, she would immediately put the name online.' As there were 49 victims on the list, publishing the names naturally put an extra mental toll on the communications team.

Facebook Safety Check

The night of the Pulse shooting was the very first time Facebook launched its Safety Check feature in the U.S. (The service was first initiated after Typhoon Ruby in the Philippines in 2014.) While Safety Check in the beginning had to be 'turned on manually' by Facebook staff, the company's algorithms turned on Safety Check automatically at 3:37 am, about one and a half hours after the shooting started (Metz, 2016).

> "That Safety Check portal became the source of news they cared about most.
>
> How Facebook Is Transforming Disaster Response, (Wired), Metz, 2016

Lessons Learned

One of the learning points for the communications team in Orlando was how swiftly focus shifts from immediate response to messages of uniting the community. A tragic event like the Pulse shooting could have torn the city apart and created hate groups and verbal attacks, but the mayor and other leaders did a great job in focusing on unity. Other lessons learned were the need for more bilingual staff and to better plan for an influx of people wanting to help.

FIGURE 3.8 At the time, the Pulse shooting was the deadliest terrorist attack in the U.S. since September 11, 2001.
Source: John Rothwell.

The City of Orlando were able to use a number of digital tools that helped them communicate effectively:

Wufoo forms were used for registering requests and incoming messages, and this data was later transferred to other databases and reports.
Password Manager Pro helped staff access applications and systems from anywhere.
Google Doc was used to share and collaborate on content and social media plans remotely.
Nextdoor, a social network for neighborhoods, proved valuable in targeting messages to specific audiences.

These learning points came to good use in June 2017, when the city marked the one-year anniversary of the Pulse shooting. The city again operated its Emergency Operations Center and ESF-14 function. They responded to media calls, monitored social media to answer questions, posted event logistics such as road closures, and thanked well-wishers from around the world. 'Multiple events took place, and the City of Orlando was able to inform the public and engage in conversations in an even better way than before the mass shooting,' Richards explained. The city's social media platforms shared content that honored and remembered the 49 victims, and the Orlando Police Department account was ready to be utilized if there were any law enforcement issues. Thankfully, there were not.

BIBLIOGRAPHY

Bindley, K. (2013, April 26). *Huffington Post*. Retrieved from www.huffingtonpost.com/2013/04/26/boston-police-twitter-marathon_n_3157472.html

Burke, S., Sims, A., & Sterman, D. (2016, October 25). *Newamerica.org*. Retrieved from www.newamerica.org/resource-security/policy-papers/war-and-tweets

Chang, K. B. (2013, April 20). Boston bombings: Social media spirals out of control. *Los Angeles Times*. Retrieved from http://articles.latimes.com/2013/apr/20/business/la-fi-boston-bombings-media-20130420

Eversley, Y. A. (2013, April 18). Obama provides comfort in Boston: 'You will run again.' *USA Today*. Retrieved from www.usatoday.com/story/news/nation/2013/04/18/boston-marathon-bombings-obama-interfaith-service/2092901

Fox, E. (2011, July 26). Norway Tribute: A nation's rose march for 'democracy,' 'unity' and those lost. *Express*. Retrieved from www.express.co.uk/news/world/261203/Norway-Tribute-A-nation-s-rose-march-for-democracy-unity-and-those-lost

Menino, T. M. (2014, April 9). *Boston Globe*. Retrieved from www.bostonglobe.com/magazine/2014/04/09/thomas-menino-view-from-sidelines-marathon/6jsHt1EIKm1IK98ZvRIYvN/story.html

Metz, C. (2016, November 10). How Facebook is transforming disaster response. *Wired*. Retrieved from www.wired.com/2016/11/facebook-disaster-response

Statoil.com (2013). Retrieved from www.statoil.com/no/wherewe-are/algeria/the-main-conclusions-of-the-investigation.html

Surowiecki, J. (2013, April 23). The wise way to crowdsource a manhunt. *The New Yorker*. Retrieved from www.newyorker.com/news/daily-comment/the-wise-way-to-crowdsource-a-manhunt

Weiss, M. (2016, January 22). Lessons from Boston's experiment with The One Fund. *Harvard Business Review*. Retrieved from https://hbr.org/2016/01/lessons-from-bostons-experiment-with-the-one-fund

Yesnowitz, J. C. (2014). The One Fund Boston: Lessons for leaders. *Boston University OpenBU*. Retrieved from https://open.bu.edu/handle/2144/8943

CHAPTER 4

Introduction and Models for Crisis Communication

Communicating with a variety of stakeholders in times of 'business as usual' is normally not especially challenging. There might be internal conflicts to handle, a negative newspaper article to respond to or a social media message that did not work as intended. In daily life, communicating is part of the routine, but those days should also be used to plan for the worst. When disaster strikes, crisis communication becomes vital. How a company or organization responds to victims, the media and the public can change the course of the crisis – for better or worse – and in some instances end careers, reduce stock prices or have a catastrophic impact on reputation.

In 'the old days' (which is not that long ago), communications teams and management had some time to agree on how to handle the situation and what to say publicly. Some textbooks talk about 'the golden hour,' explaining that the first hour after an incident could be used for planning and getting ready. With social media, such a 'luxury' is no longer part of the scenario, and more often than not news about a tragedy appears first on Facebook or Twitter. Soon afterwards, the public and the media expect answers, and the whole world can scrutinize how you respond and what your priorities are.

If you do not provide an immediate response, you are seen as hiding and being evasive. This creates a vacuum others will fill, and it is therefore good practice to make your presence as a communicator early – as early as possible. By posting a Facebook message or tweeting a few words shortly after something has happened, you demonstrate that you and your company will be a voice to be reckoned with.

The National Transportation Safety Board usually tweets about an accident within an hour, thereby showing the public, the media and airlines that an authoritative voice is already on the case.

The first message does not need to be very long. It could suffice to:

- Acknowledge what has happened
- Explain what is being done
- Show empathy
- Provide a link to more information

Transparency, openness and care are words that are often used to describe the goals of crisis communication. They are not easy to achieve, and in a crisis situation there can be a tendency to avoid giving out bad news in the hope that it will not otherwise surface. That is very seldom the outcome, and if a newspaper or a citizen instead is the source providing the negative information, the company or organization often finds itself in trouble. It is therefore wise to be open and give out all kinds of news, even if it involves admitting a wrong decision or having provided the incorrect facts.

Care should be a vital word in all crisis communication. A crisis always involves victims, and when communicating it is imperative that the focus is on those who have been hurt or those who have lost loved ones. As described in Chapter 10, families should always be informed before the media, which can be accomplished through a Family Assistance Center or through a dedicated web page for victims. At the same time, though, it is vital that as many people as possible see themselves as survivors instead of victims of a disaster, a point President Obama described in his commencement address at Joplin High School in 2012: 'We can define our lives not by what happens to us, but by how we respond. We can choose to carry on. We can choose to make a difference in the world' (Obama, 2012).

MODELS FOR CRISIS COMMUNICATION COLLABORATION

Getting messages across in a timely, compassionate and understandable way takes planning and collaboration between several parts of an organization.

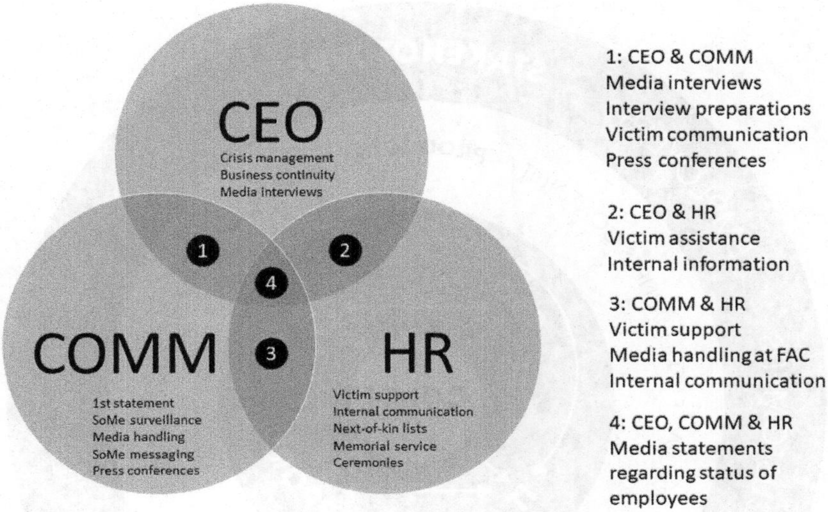

FIGURE 4.1 'The 4C Model' *Circles of Crisis Communication Collaboration* describes how the CEO, the communications team (COMM) and human resources (HR) need to collaborate in times of crisis.
Source: Kjell Brataas.

As the 'Circles of Crisis Communication Collaboration' model in Figure 4.1 describes, three parts of an organization need to work exceptionally well together in times of crisis. The CEO, of course, takes charge and makes priorities, but she or he needs a lot of advice from HR and Communications. The CEO and COMM, for example, collaborate on media statements and press conferences, while the CEO and HR will work together on victim assistance.

The second model of a 'Communication Product Loop' (Figure 4.2) explains how even more parts of an organization interact in order to produce communication 'products' (press releases, messages on social media, web pages etc.) for various stakeholders. This model also includes IT (handling computer access etc.) and a support group or secretarial service (handling a variety of tasks such as copying, food and drinks, transportation requests etc.).

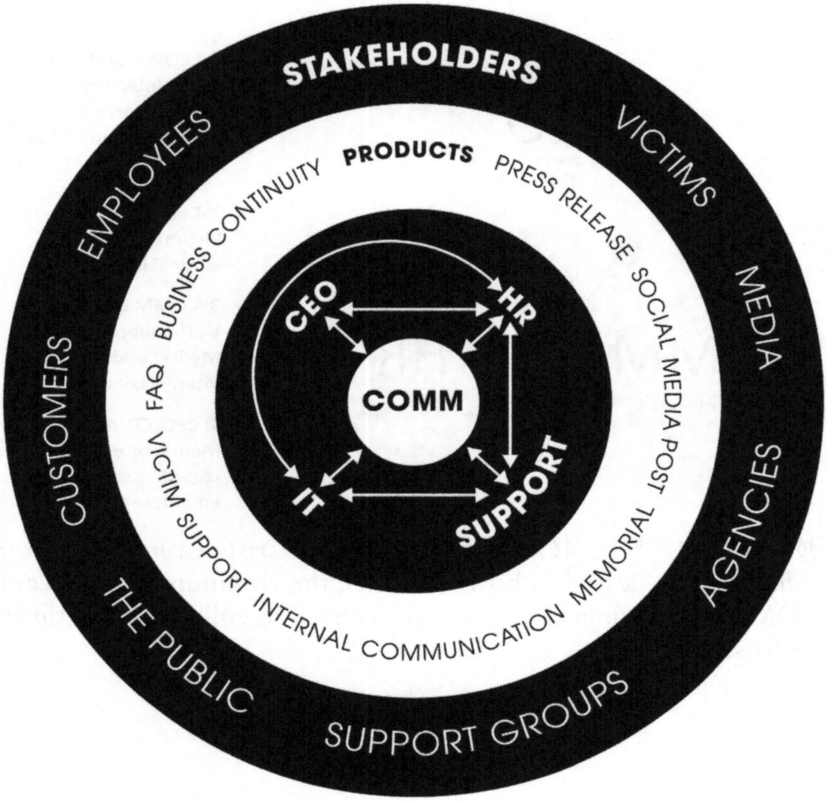

FIGURE 4.2 The 'Communication Product Loop' can be expanded or changed based on type of organization, crisis and so on.
Source: Kjell Brataas.

BIBLIOGRAPHY

Obama, P. B. (2012). Read President Obama's Commencement Address at Joplin High School. *Time*. Retrieved from http://time.com/4341639/obama-commencement-speech-transcript-joplin-high-school

CHAPTER 5

Working with the Media

Even with a large part of the world using social media on a daily basis, traditional media should not be forgotten when planning for or handling a crisis. People still turn to local news or CNN when something has happened, and press releases and one-on-one interviews continue to be an effective way of getting messages across. What is new, however, is that more and more people consume traditional news sources and social media simultaneously. Especially after a tragedy, people seeking information will be watching the news on TV while at the same time following Twitter feeds and checking Facebook. The sources of news reports have also changed dramatically, as more and more pictures and videos from the scene of an accident come from bystanders with smartphones.

In addition to being a news source, traditional media have an advantage in also offering in-depth analysis, critical voices, impartial studies and commentaries. Although not always positive, it is important to have these aspects in mind when planning for and dealing with the media, and to have systems in place for media surveillance.

Even in a stressful crisis situation, and with journalists uncovering unflattering news or asking questions that are hard to answer, we should remind ourselves that having a free press is part of democracy. Victims, citizens and the public in general deserve to know the truth, even if it is sometimes not favorable from our standpoint. Getting facts straight and admitting mistakes can contribute to making sure the same tragedy does not happen again.

FIGURE 5.1 Media interest can escalate quickly, as was the case after the terror attack in Oslo in the summer of 2011.
Source: Kjell Brataas.

ACCEPTING AND ANSWERING MEDIA CALLS

The press office of a private company, a government ministry or a hospital might receive hundreds of phone calls from journalists in a few hours after a tragic event. Such calls create at least two imminent questions that need to be planned for internally:

- How do we handle and process the calls?
- What do we answer?

The first question concerns preparation (see also Chapter 11), as plans for taking calls from journalists should be in place before something happens. As a minimum, there should be a dedicated phone number and a technical system that allows several people to answer. There must also be plans for how to process incoming phone calls and their follow-up, which is often accomplished by using an electronic logging system. Spreadsheets like Excel can do the trick, but it is often better to set up a simple, web-based system that several people can access at the same time (Google Sheet is a free and simple solution to consider).

Shortly after an event, there is usually not much concrete information to give out. Incoming calls from journalists can therefore be answered with basic facts about what has happened and some words on what is being done about the situation. The focus ought to be on the victims and immediate support and response, and questions about cause or blame should not be answered at this time. If a press conference is scheduled later in the day, it is a good idea to refer journalists to that event and to publicize an invitation to the media on a website and in Twitter messages. Knowing that there will be a press conference later can lessen the burden on the press office staff, who instead can focus on preparing and checking facts and developing the right messages for later dissemination.

If the crisis being handled has an international aspect, for example if it involves foreigners or has happened in another country, the communications staff must be prepared to handle phone calls from journalists speaking several languages. If we are talking about an aircraft accident, the airline should expect phone calls from all the countries that had passengers on the flight.

Journalists calling a press office often have deadlines. While they are not always possible to honor, the communications team should try to get answers or at least call back within the deadline specified. If a journalist asks for a live interview with the CEO, this should be viewed as an opportunity to get your side of the story in front of a large audience.

THE PRESS RELEASE

Press releases should be short and to the point. They usually follow the style of an upside-down pyramid, with the most important message at the top followed by a short statement and additional facts and information. The press release must convey sincerity and compassion and focus first and foremost on the victims and on what is being done to lessen the burden.

A press release needs someone to receive it, preferably news organizations, journalists and editors. How it is distributed should be planned before a crisis; it can be in the form of an email distribution list or a paid service through a third party provider.

Bear in mind that the press release is not written for the press alone. It should therefore be published on the organization's own web page, be linked to from a message on Twitter and have some of its contents used in a Facebook post.

PRESS CONFERENCES

Organizing a press conference can have several merits:

- It provides time to prepare and get information and facts straight.
- It makes it possible to discuss messages and key points with other parties involved.
- It gives the communications staff a break from a bombardment of requests for interviews.
- It is a concrete example of something being done and that you are 'on' the case.
- It is an opportunity to show leadership and commitment to resolving the issue.

A traditional press conference should be combined with social media, for example by having a live video feed on Facebook or by live-tweeting key points. You could also videotape the whole press conference and post the recording on your YouTube account.

A press conference usually follows the following structure:

1. Introduction.
2. Head of company/CEO/mayor etc. talks about what has happened and what is being done.
3. Someone else, such as an expert or the police, says something and gives more details.
4. Journalists can ask questions.
5. The formal part of the press conference ends, and journalists can have one-on-one interviews.

It is important to set aside enough time for one-on-one interviews, as these tend to be popular and take time. The person being interviewed should be told to stick to the points outlined for the press conference, and never to break news in a one-on-one setting.

To be mutually effective, press conferences need firm leadership. A senior communications staff member usually functions as the moderator and starts by giving a short summary of the order of events before presenting the panel. Journalists wanting to ask questions should be told they have to raise their hand, and when they are called, state their name and which media outlet they represent. If the press conference is being recorded or broadcast live, everyone should use a microphone. The moderator also plays an important role in keeping order and making sure the conference does not last too long. If someone repeats a question, the moderator should immediately jump in and say: 'That's been answered, let's move on to NBC.'

If the press conference involves a tragic event such as an airplane accident, an appropriate start could be to have all participants stand while one person reads a joint statement about what a tragic day it is, that they are all sorry and that they are doing everything they can to help.

As with all kinds of crisis communication, preparation is essential. Before a press conference, the CEO or other spokesperson should be briefed about the kind of questions that might come up and appropriate responses. Questions might be quite personal, and the person in front of the microphone needs to be prepared when journalists start asking 'impertinent' questions such as 'Will you resign?' Tell the spokesperson that these types of questions are part of the business and what viewers want answered – and that losing one's temper on camera makes for great TV …

In a very few instances, a next-of-kin might show up at a press conference and start asking questions. This should, of course, be avoided if possible, as such an appearance is likely to harm the family member in the long run and totally take away from the messages being conveyed from the podium. This is another reason for having an arrangement in place that makes sure that only journalists are present at the press conference, and that security is available in case someone else tries to 'break in.' If a relative of a deceased or missing person nevertheless enters the arena and takes the spotlight, the CEO or moderator should stop the press conference and walk up to the relative. A possible approach could be to say gently: 'Would it be OK if we speak in private in a few minutes, just you and I?' and then suggest that he or she probably does not want to be on the evening news.

PREPARING FOR AN INTERVIEW

Below are seven questions worth considering before giving an interview:

1. What is the setting?

Find out where the interview will take place, who the journalist is and who else has been interviewed about the same story.

2. How to prepare?

Do your homework. Know as much as possible about what has happened, the order of events and who is handling the situation. Looking at other interviews conducted by the same journalist might give some ideas about the kind of questions that will be asked and the 'tone of voice' that can be expected. This can also be accomplished by checking the journalist's Twitter profile. (Ideally, all CEOs and prospective spokespersons should go through media training before a crisis.)

3. What if you don't know the answer?

No one is expected to have all the answers to questions that can arise after a tragic event, especially not in its early phases. If you do not have a concrete answer, it is OK to say that in an interview – ideally in combination with a 'bridging technique' (described below): 'I'm sorry, I don't know. However, what I can say is ...' This is especially important if the journalist wants details about numbers. Offer to get the facts checked, then get your communications team to find the answers – and get back to the journalist as soon as you have them.

4. How to 'steer' the interview?

While you should always answer the question the journalist poses, do not be afraid to take the initiative and talk about what you want to say. One technique to use is called 'bridging,' which means that you – as eloquently as possible – move from one sentence to an answer that is more in line with your key messages. For this to work, you need to keep talking and say something like '... but what is important for us now is that ...' or '... at the same time, I want to point out that we are ...' Another option is to mention something in your answer that will make the journalist ask for more, for example: '… that is not what is most important for me right now.'

5. What if I or we have done something wrong?

If you personally, your staff, the company or organization has done something wrong, the right thing to do is to apologize. Use simple words and few sentences and convey sincerity when admitting to having done something wrong. It might be a hard thing to do, but the future will look much brighter if you admit at once, instead of trying to cover up, blame someone else or dodge the question.

6. How should I handle 'small talk' before and after the interview?

You should consider all conversations with a journalist as part of the interview. Do not be tempted to 'confess' while off camera or to say something that is contrary to your main points.

7. How should I answer questions about competitors, other agencies and so on?

While it might seem tempting, avoid saying anything negative – and say as little as possible – about your competitors. The interview you are giving will likely get only a few seconds of air time; use it wisely to talk about your organization and your messages.

WHEN THE RED LIGHT FLASHES – THE INTERVIEW SITUATION

Giving an interview in the early phases of a crisis can be trying, but for the first few hours most questions will concern what has happened and what is being done about the situation. Nevertheless, preparation is the key, and a useful approach is to coin two to three sentences that say the most about the situation – from your point of view. This does not mean you have to keep to a script, and when the interview starts it is OK to be personal. If you say something like 'This is the worst day of my life,' or 'As a father myself, I can only imagine …,' you come across as someone who takes the tragedy personally and as someone with compassion and a human side to a company or an organization.

While the person being interviewed should try to convey his or her predefined messages, it is important not to totally ignore the journalist's questions. If you never answer the question and talk about something else, you instead appear as a person having something to hide, and you will make the journalist as well as the viewers angry.

Take this example from an interview on 'Good Morning Britain' in June 2017, during which journalist Piers Morgan tried to get an answer from Culture Secretary Karen Bradley to a simple question about the number of armed police (MediaFirst, 2017):

Piers Morgan: Do you know if the number of armed police has gone up or down in the last six years? Do you know the answer?

Karen Bradley: Piers, what I am interested in is making sure that we have the right resources, the right powers and the right training and capabilities. I am assured by the police that they have that to deal with the counter-terrorism threat, but we need to look, learn lessons and make sure we act where appropriate ...

Piers Morgan: Is there any reason why you can't answer the question?

Karen Bradley: Piers, we are here to talk about how we react to the attack on Saturday.

Piers Morgan: So, you just don't want to answer?

Karen Bradley: Piers, we are here because on Thursday there is a general election.

The exchange made Ms Bradley seem robotic and scripted, and social media reactions on the same day were harsh in their criticism, like this message on Twitter: 'Ceiling-shattering levels of evasiveness, even for a politician ...' (MediaFirst, 2017).

Some other advice for the interview situation:

- Do not speculate.
- Do not repeat exact wording from a press release.
- Do not repeat negative phrases used by the journalist.
- Do not predict the future or answer hypothetical questions.
- Do not give guarantees.
- Do not give different answers to the same (repeated) question.
- Do not fill 'dead space' during an interview with additional information.
- Do not be afraid to repeat your talking points.

OTHER CRISIS COMMUNICATION PRODUCTS

Media Surveillance Reports

In the early phase of a crisis, people reporting on social media and traditional news outlets often know more about what is happening than the organization handling it. Bringing news and rumors to the attention of senior management, a media surveillance report can therefore be a valuable tool for assessing what is happening and what needs to be done. Producing such a report takes knowledge and experience and should therefore ideally be handled by a senior communications expert. The report should include bullet points about what is being reported by news organizations and on social media (with links and screen clippings) and also sentiment evaluation and a few words on how the overall media picture might develop.

Photo Bank

When crisis hits, communications staff do not want to be caught up in mundane tasks of providing the media with pictures of CEOs or their office building. To avoid this, every company or organization should have a web page dedicated to the media, where pictures of management and other generic images can be downloaded in high resolution. Specialized web services are also available, such as Flickr or Google Photos.

Video Statements

In addition to appearing on TV news, the head of the company handling a crisis could give a statement on a social media channel. Many organizations have an established presence on YouTube, and a short message from the top manager can work well in conveying sincerity and management involvement. The video could be recorded and edited on a smartphone, but it is important to use a tripod and a microphone, which gives the result an acceptable quality. The finished video should also be published on the company website and social media.

BIBLIOGRAPHY

MediaFirst. (2017, June 5). *Mediafirst*. Retrieved from www.mediafirst.co.uk/our-thinking/four-quick-lessons-from-this-evasive-interview

CHAPTER 6

Social Media in Crisis Communication

A BRIEF HISTORY

As will be clear from the case studies and other chapters presented in this book, social media have dramatically changed crisis communication and the way it is conducted. There is no longer a discussion about *whether* or *when* to use social media, but *how*! The focus is on how to use and incorporate social media into daily routines, how to improve virtual collaboration, and how organizations and top management can succeed in crisis communication – with the help of social media.

The watershed moments for social media and crisis communication came with the earthquake in Haiti in 2010 and Hurricane Sandy in 2012. After the earthquake, traditional telecommunications infrastructure crumbled, making Haitians in Port-au-Prince as well as the Haitian community in Miami turn to social media, and especially Twitter. Initially focused on information and reports about what was happening, tweets soon turned to requests for aid and donations. Many news organizations based their reporting on social media, especially CNN, which utilized its iReport section of its web pages for people uploading video from Haiti. They also published articles which mostly contained information from Twitter, such as a story titled 'What we're hearing via social media' (CNN, 2010).

The use of social media became even more prevalent during and after Hurricane Sandy in 2012, and Sandy was one of the first examples of government agencies using several social

FIGURE 6.1 Governor Christie and other leaders used Twitter to reinforce important messages.
Source: Twitter.

media platforms to communicate with the public and maintain awareness in connection with a major disaster. It helped that New York City was *ready*. Before the storm hit, the city already had 300 different accounts on Facebook, Twitter, Tumblr and other channels (Cohen, 2013). The city monitored social media for reactions to the storm and used data collected to send reports to City Hall, while government officials with thousands of followers used social media to actively spread vital information about the storm. The potential was much greater, though, as of the 840 fire and police departments within a 100 mile radius of Sandy's landfall, only 7% used Twitter and 25% used Facebook (Hughes, 2014).

On a side note, it is tempting to think of how another disaster – Hurricane Katrina in New Orleans in 2005 – would have played out if social media had been around at the time. There is no doubt that the communication part of the crisis would have been very different, but some commentators believe lives could have been saved through Facebook, Twitter and YouTube. In an article in *Washington Post*, Jason Samenow argued that 'In today's social media climate, the echo chamber about how grave a threat Katrina posed would have been deafening. The hype would have appropriately blown through the roof and preparedness messages would have spread like wildfire' (Samenow, 2015).

BENEFITS

Fast-forwarding to today, it is easy to see a number of reasons for using social media during a crisis:

- The public expect it.
- The tools are often for free.
- News often breaks on social media.
- Social media enable dialogue.
- Social media function as an alternative channel when normal channels cannot be used.
- Messages can be targeted to specific groups of people.
- Many social media channels offer advanced analytics.
- Social media can enable crowdsourcing.
- Messages on social media often require fewer steps of approval.
- Social media can be used to give instructions to people who might be harmed.
- Social media can help humanize an organization or government agency.
- Through social media, the public can be supplementary 'eyes and ears,' thereby providing an additional situational awareness tool.
- Social media can be used to promote volunteer possibilities.
- Social media can be used to kill myths and rumors.
- Social media staff can be 'distributed'; they can work from locations other than main office, from abroad or from a café.
- Social media can act as free and easy tools for sharing 'safe and well' messages.
- Event detection is possible through social media.

> "We can have conversations with the citizens, and they with one another, in a public forum for all to see. Through this type of dialogue, you start to understand your community and what is important to them. That is invaluable.
>
> Mastering Social Media: An Analysis of Jefferson County's Communications during the 2013 Colorado Floods (St. Denis, 2013)

Although traditional websites of government agencies, organizations and corporations are still central in a crisis situation, social media have benefits that can amplify important messages. One obvious and important advantage is that social media can be an alternative channel if a website crashes or has other technical issues. After a tragic event, traffic to a website can escalate tremendously, often resulting in no one being able to reach the important information published there. Facebook, Twitter and other social media channels can then provide the same information.

Traditional web pages often require a log-on procedure involving usernames and passwords, and many websites are best updated from a PC. Social media channels are easier to connect with, and they enable communications staff to publish 'on the go' from a smartphone or a tablet.

> "
> Using social media as part of a comprehensive approach enabled CEMA and the City of Calgary to lead the conversation from the start, communicate reassurance, and explain decision-making to the public.
>
> Calgary Flood Report (Vroegop, 2014)

ESTABLISHING A PRESENCE

When you register a username on Twitter or an organization's presence on Facebook, you start with zero followers. To be an effective tool, social media therefore need to be thought of *before* something bad happens and to be part of everyday life in the office. Routines must be in place for handling several administrators and editors, and information about suitable picture sizes and how to upload images should be easily available. Another useful routine is to have a dynamic social media calendar, which is populated with important events and historical notes that can be used for publishing on Facebook, Twitter or other social media channels. Also, there need to be staff in place and internal guidelines in order to decide what kind of information should be distributed through the various channels and platforms.

Some other best practices:

- The profile photo is noticed more than you might think, and it is therefore a good idea to modify it regularly. If a crisis occurs, consider promptly changing it to a neutral image of just a grey or black color.
- No one expects an organization or a company handling a crisis to answer every question on social media. However, those questions convey important clues about what the public is thinking of and what their focus is. You should therefore sum up some of them and give answers in more general terms, and use the same information for an FAQ on the official web page.
- Long form content can also have a place on social media, for example in the form of articles or background information posted on Facebook Notes or similar.
- Social media monitoring can be used to identify reputational threats and to gage how your messages are being perceived by the public.

POLICING THROUGH SOCIAL MEDIA

Police forces in many parts of the world profit from having an established presence on social media, in everyday life as well as in times of crisis. When the bombs went off near the finish line of the Boston Marathon in 2013, the Boston Police benefitted greatly from their #TweetfromtheBeat Twitter interactions with citizens since 2011, which helped ensure the public that the police Twitter account was one they could trust. Several police forces also utilize social media to show a lighter side of their work. One example is the Bangor Police, whose public information officers invoke humor in a way that attracts followers from far outside their jurisdiction. They use Facebook for branding the Bangor Police and do not shy away from long posts without pictures. One especially popular post suggested to local residents how they should behave if interviewed by the media about winter storms. It was shared more than 5,000 times and combined serious preparation advice with hints such as

> While being interviewed, make sure there is a shovel somewhere in the camera shot. People want to believe we always have a shovel.

> This in turn makes them feel badly for us, and when they return in the summer they will tip better at our local restaurants.

After several similar jokes, there followed sections with 'real' messages, as in: 'On a serious note, make sure that your woodbox is filled, generators are serviced and ready, fuel tanks have plenty with some reserve, and anything that can blow away is secured' (Bangor Maine Police Department, 2016).

Another police force that has been in the forefront of social media is the police in Iceland. They experimented early with lost and found pictures on Pinterest and 'tweetathons' that included descriptions of all cases the police worked on for a 12-hour period. 'We view our social media channels as flowers in a garden. Daily care and frequent updates make activity grow,' explained Thorir Ingvarsson, who headed the Iceland police social media presence. He says that by communicating on an everyday basis, the police build a relationship that can be used to help them in a crisis situation. Ingvarsson explained further:

> Social media is a way for people to get messages straight to you, meaning that when people are dissatisfied with the service you are delivering they will tell you straight away. I think of this as a contract of sorts where you agree that if you can have a direct line to the public you serve, they in turn have a direct line to you. That means that a complaint might be posted on your wall where everyone can see it and we have to deal with it publicly. In my opinion, this is not negative, but it can be challenging, for obvious reasons. This also means that when people contact you directly, they are in fact giving you a possibility to respond and fix the problem. If you do just that, people will usually be very happy and satisfied that you took the time to remedy their situation.

RULES OF ENGAGEMENT

It is a good idea to post 'rules' on Facebook and other platforms regarding what kind of comments will be accepted, for example in the form of a few sentences:

> We welcome your comments and invite you to share your ideas on our Facebook page. We ask that you do so without using foul

language, sexist remarks or otherwise inappropriate words. Please be 'on topic' and do not threaten or defame any person or organization. Failure to follow these guidelines might result in posts being deleted or individuals being banned from commenting.

Having guidelines in place makes it easier for the communications team to assess whether comments are acceptable or not, and it helps to have concrete rules to refer to when informing individuals that their comments will be deleted. (It is always a good practice to keep a screen shot of the offending post on file, in case the person submitting it or others start debating the 'censorship.')

Removing offending messages should be a last resort, and crisis management as well as the CEO should be told that you will allow negative comments and open criticism. Removing a negative comment can easily result in a blowback, and as long as comments conform to your posted guidelines, they should be allowed.

SOCIAL MEDIA MONITORING FOR FACTS, RUMORS AND FAKE NEWS

Many case studies and previous events show the importance of monitoring what customers, journalists, competitors or individuals say about your company or industry. This should be done on a routine basis, but in times of crisis it needs to be escalated to include more key words or hashtags. Fortunately, there are several tools available; some of them are even free.

In a crisis, citizens and victims often become reporters, giving those handling the situation an excellent tool for situational awareness and on-site intelligence. These data are sometimes more correct than what is being reported by the police or in the media, as was the case in Norway in the evening of July 22, 2011, when at 9:38 the police confirmed ten deaths on the island of Utoya. Shortly afterwards, an individual who knew a survivor tweeted: 'I have a friend who barely escaped Utoya. He is convinced the number of deaths is much greater than the 10 confirmed.' (The correct number was 69.)

Monitoring also includes picking up concrete questions and requests for help through social media, as described in a research paper from the University of Colorado, Boulder: 'The public expects a group with a Twitter presence to monitor their questions and concerns, and to respond' (St. Denis, 2013). Requests for emergency assistance *will* appear on social media, and it does not help to have a statement saying that such requests must be made through a telephone call to 911 or similar services. During a crisis, often telephone lines are down, and citizens in despair will then ask for help on one or more social media channels, and they expect to be heard.

After a large tragedy, misinformation and fake news often appear on social media. There are several reasons for these postings; some come from people who do not know they are spreading false information, while others deliberately publish messages to confuse others or to hinder investigation. While you will not be able to *control* rumors, it is vital to follow conversations closely and to have a system in place for correction, as the Queensland police did effectively during the flooding in 2013 (described in Chapter 2).

It is important for everyone in crisis management positions to realize that volume does not equal facts when it comes to posts on social media. Even if many people give out the same information or images, it does not mean they are conveying the truth. Having an authoritative voice and a constant presence on social media can help, and the police or city officials can avoid confusion by telling citizens that *their* information is correct, and that they should turn to these channels for confirmed facts. Another solution is to set up a dedicated website for 'rumor control,' as FEMA did after Hurricane Sandy and Hurricane Irma.

Verifying content on social media can be done in a variety of ways. A study titled 'Information Verification during Natural Disasters' suggests considering the number of Twitter followers a user has and whether the Twitter account is followed by reputable sources, and checking how long the Twitter account has been active. Other verification methods regarding pictures can include considering time of day of pictures (shadows etc.), weather and landmarks (Popoola, 2013).

A useful tool for verification purposes is Google's Reverse Image Search. It enables anyone to search by images instead of keywords, and the result is a list of sites that include the image as well as visually similar images.

> "Twitter is for delivering the news. Facebook is where we talk about the news, and the blog is where we provide the details.
>
> The Jefferson County IMT Team (St. Denis, 2013)

An organization that used social media successfully during a crisis is the Jefferson County Sheriff's Department and their Incident Management Team (IMT). When a large flood occurred in 2012, they used what they call an 'integrated social media strategy' that included social media surveillance and use of digital volunteers (VOST) to monitor and amplify messages. According to the report 'Mastering Social Media: An Analysis of Jefferson County's Communications during the 2013 Colorado Floods' (St. Denis, 2013), the team used multiple strategies for getting messages across a variety of platforms. Relevant hashtags, frequent messaging and good hyperlinks were key reasons for their success. By providing a simple Google form for uploading pictures and video, they could ask the public for help with situational awareness, and with the assistance of a VOST team, they were able to squash fraudulent tweets asking for donations.

Fake Content from Boston

According to research from 2013, a lot of fake content and malicious profiles originated on Twitter after the Boston Marathon bombings: 29% of the most viral content on Twitter during the Boston crisis was rumors and fake content, while 51% was generic opinions and comments, and the rest was true information. According to the report: 'Many malicious accounts were created on Twitter during the Boston event, that were later suspended by Twitter. We identified over six thousand such user profiles' (Gupta, 2013).

> “
> Social media can act as excellent tools for situational awareness and intel gathering.
>
> Christopher Tarantino, CEO of Epicenter Media & Training (personal communication)

SOCIAL MEDIA MESSAGES

Twitter, Facebook and other social media channels are excellent tools for giving instructions when a crisis has occurred. There are many good examples, such as the Belgian authorities using Twitter for warning the public not to go the airport after the terror attack in 2014, or German police urging people not to come to the site of the bombing in Berlin as they needed as many routes as possible for police cars and ambulances.

The speed at which an organization handles social media challenges – and releases information – can be vital to the outcome of a crisis. Crisis managers often want to wait until information is confirmed and fact-checked, but for greatest effect, the first messages need to appear shortly after a crisis has occurred. A simple confirmation from an official source can suffice, such as a short Twitter message stating: ‘We are aware of the situation in India and will publish more information shortly.’ Such a message shows the public that your organization will be a voice to be reckoned with, and that you are ‘on the ball.’ In fact, research shows that messages of official communicators can rise above the noise of others and that engaging in the online conversation can dampen the spread of rumors (Andrews, 2016). To be able to distribute the first messages quickly, it is therefore a good idea to have templates of statements available that do not need managerial approval. A best practice in this regard is the Canadian airline WestJet, which has 100 precrafted ‘stock’ tweets, approved by the executive team, designed for various events (Andrews, 2016).

Some messages on social media can be generic, and it is a good idea to have a ‘bank’ of instructions thought of in advance of a crisis (and part of the crisis communication plan). A few examples:

- 'Do not spread rumors or images of victims.'
- 'Out of #respect for victims, please restrain from publishing or sharing any sensitive images.'
- 'Follow local authorities and reliable sources for advice and information.'
- 'Use SMS instead of calling.'

In addition to being a source of information, social media are often used by the public to help each other or offer assistance after a tragic event. Twitter can be used effectively for connecting with strangers, and hashtags such as #openhouse or #RoomForManchester have proven valuable when stranded individuals need a place to stay.

There are many examples of social media bringing a *personal* side of crisis communication into the realm of the public. The CEO of AirAsia, Tony Fernandes, was praised for his use of Twitter after flight QZ8501 crashed in 2014, as his tweets rallied his team, provided support and gave details about the tragedy. Individual responders also have a voice through social media, and in June 2017 an English firefighter tackling the Grenfell Tower blaze received worldwide attention after tweeting a photo of his helmet and a simple message: 'You know it's not going to be good when you are told to write your name on your helmet before you go in!'

The main reason for publishing a message on Twitter is for people to read it, and a useful tool for reaching a large audience is to use relevant hashtags like #jeffcoflood or #Manchester. Hashtags make searching more effective and allow people to follow messages from a variety of sources regarding one event. Through tools for social media management, journalists and citizens alike can easily organize hashtagged messages on Twitter about an emergency, and they should therefore include at least one hashtag. (A world record of hashtag use might have been set by the social media teams communicating about the flood in Calgary in 2013. According to a report by the Conference Board of Canada, the hashtag #yycflood showed up an average of 32 times *every minute* over a 10-day period; Vroegop, 2014.)

Standardizing Hashtags

Over the years there have been several discussions about best practices and suggested standards for the use of hashtags. In 2014, United Nations Office of the Coordination of Humanitarian Affairs (OCHA) published a study that came up with three types of hashtags to be used during a crisis; a disaster name hashtag, a public reporting hashtag and an emergency response hashtag.

LIVE REPORTING THROUGH PICTURES AND VIDEO

It is nothing new that individuals can report directly from a tragic event to the police and to the media. Ever since their widespread adoption in the 1990s, cell phones have proved a valuable communication tool in disasters, and a 1999 Library of Congress report on the 9/11 attacks mentioned that 'the proliferation of cellular phones in recent years has made it possible for emergency victims to interact directly with the media.' Another development came around 2000, when camera phones became popular. Their importance was first demonstrated in July 2005, when commuters provided pictures from the bombing of the London metro system. According to an article in *National Geographic*, cell phone video from the area was airing on television within 30 minutes of the bombing.

The change with *social* media is that such footage now can have a global audience – and be broadcast live. Unfortunately, it is not only the victims or the bystanders who can report live on social media, as the perpetrator has the same ability. There have been gruesome images posted by shooters or terrorists on Facebook, Twitter, Instagram and Snapchat, and such posts can unfortunately spread fear and despair to a large audience.

There are advantages too, of course, and live reports can be of excellent help to anyone in law enforcement or crisis management whose task it is to figure out what is happening.

DISASTER RESPONSE THROUGH FACEBOOK

In 2014, Facebook launched a new feature they called 'Safety Check.' The company realized that people were using Facebook to tell their friends and family members they were OK after a crisis, and Safety Check became an easy-to-use and available social infrastructure to help communities connect. The CEO of Facebook, Mark Zuckerberg, explained the reasoning behind Safety Check in a post on Facebook on October 16, 2014: 'Over the last few years there have been many disasters and crises where people have turned to the Internet for help. Each time, we see people use Facebook to check on their loved ones and see if they're safe. Connecting with people is always valuable, but these are the moments when it matters most. Safety Check is our way of helping our community during natural disasters and give you an easy and simple way to say you're safe and check on all your friends and family in one place.'

The earthquake and tsunami in Japan in 2011 was the event that created an interest within Facebook to develop a new tool for connecting people after a disaster. Their engineers in Japan took a first step by building a message board, and when a test showed an overwhelmingly positive response, the project later evolved into Safety Check.

'Community Help,' an additional part of Safety Check, was launched in 2017. This feature makes it easier for people to find and give help, enabling individuals to offer shelter, seek food or offer assistance.

DIGITAL VOLUNTEERS AND THE CONCEPT OF VOST

For most of history, assisting in a disaster has been restricted to those who can physically take part, issue a check, or give money to the Red Cross or similar organizations. With social media, however, anyone can be a volunteer, making it possible to contribute from a home in another part of the world or an office in a different time zone. Surge support is now international.

FIGURE 6.2 Facebook has developed several tools for use in a crisis, including 'Safety Check' and 'Community Help.'
Source: Facebook.

Several organizations offer digital volunteers assisting in disasters. Humanity Road, Geeks Without Borders, OpenStreetMap and VOST (Virtual Operations Support Teams) are various groups that provide support, but VOST is the group with the closest ties to emergency managers.

The concept of VOST originated in the U.S. in 2010 after attendees at a national emergency conference discussed the possibility of providing 'digital' assistance in case of disasters. By then, social media were becoming a challenge, especially in regard to social media surveillance and rumors being spread on Twitter and Facebook. Individuals such as Jeff Philips, Scott Reuter and Cheryl Bledsoe in the U.S. and Caroline Milligan in New Zealand therefore started forming groups of volunteers who – with training and approval processes – could provide valuable insight into what was being discussed on social media. The idea spread to many states in the U.S. and across borders, and by summer 2017 there were VOST groups in 32 U.S. states and 18 different countries.

VOST members provide innovative resources and can assist with translating 'noise' on social media into actionable intelligence.

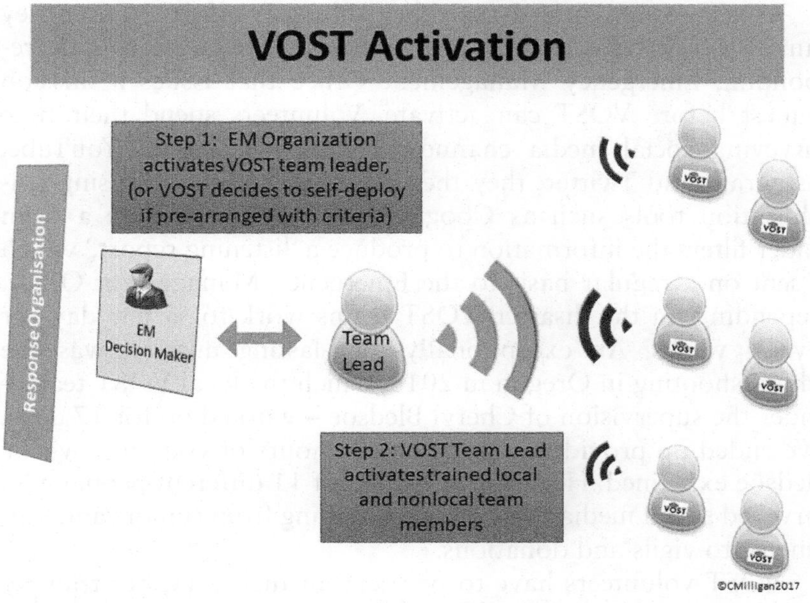

FIGURE 6.3 VOST teams are 'trusted agents' who can provide valuable situational awareness through surveillance of social media.
Source: Caroline Milligan.

They are able to validate what is being said and commented on, contextualize media reports, and follow and analyze messages on social media, blog posts, community forums and online news comments sections. Other important goals of a VOST team are to amplify official messages, detect trends and issues early, and correct rumors and misinformation.

When a large forest fire broke out in Fort McMurray in May 2016, Canada VOST received a formal request for support from the Alberta Emergency Management Agency (AEMA). The team assisted with social media monitoring and provided twice-daily reports on trends and rumors to officials handling the wildfire. 'We paid special attention to progress of evacuation and sentiments at reception centers,' said Patrice Cloutier, who headed the CanVOST team. They got valuable support from other VOST teams in the U.S. and Australia.

VOST groups differ from other volunteer efforts in that they cannot self-activate. In the U.S., each team hooks up to a corresponding Emergency Management Office that issues a mission request before VOST can activate. Volunteers spend their time surveying social media channels such as Facebook, YouTube, Instagram and Twitter; they then report their findings using collaboration tools such as Google Docs or Slack before a team leader filters the information to produce a 'listening report,' which is sent on a regular basis to the Emergency Management Office. Depending on the disaster, VOST teams work for a few days or several weeks. An exceptionally long-lasting mission was the school shooting in Oregon in 2016, which the local VOST team – under the supervision of Cheryl Bledsoe – worked on for 17 days. 'We ended up providing a total of 287 hours of volunteer work,' Bledsoe explained. Her team consisted of 11 different people who surveyed social media messages on anything from rumors and sentiments to vigils and donations.

VOST volunteers have to be social media savvy, but training and documentation are available for those who want to learn. Quite a few volunteers are 'regular people' who want to contribute by spending a few hours in front of a computer, but more and more VOST teams are being populated by professionals within the field of emergency management. With training and policy development, VOST teams have been able to also include 'traditional' practitioners and upscale them to the digital environment of social media surveillance.

The 'beauty' of VOST is that groups are scattered around the world, and with social media and digital tools such as Google Docs, they are able to work on the same issues and in the same document from anywhere. When team members in Canada or the U.S. get tired, they can ask a VOST team in New Zealand or Europe to support them and continue their work, which has proven a valuable approach in many disasters. International VOST teams also bring a non-local perspective, such as when Caroline Milligan – situated in New Zealand – functions as a team leader for Colorado VOST or Oregon VOST.

Many officers of emergency management have learned to value digital volunteers. They are seldom able to sort out and be on top of social media messages by themselves, and Director Andrew Phelps of the Oregon Office of Emergency Management points out that 'having these folks on the virtual operations support team to really get what the sentiment is for the public that we're here to protect makes that go a little bit easier for us.' His colleague Cory Grogan agrees: 'VOST works. It's needed. It's an important tool for emergency managers' (Eiten, 2015).

VOST in Colorado and France

According to the web page of the *Colorado Virtual Operations Support Team* (COVOST), their mission is to provide the State of Colorado and local Colorado communities with a powerful and effective team of trusted agents. They will 'support public information, operations, and information gathering efforts through an innovatively effective, efficient and elegant use of social media and other online or "virtual" tools.'

The French VOST team – called VISOV – mainly supports the Civil Defense Department (COGIC) in France. VISOV has also supported the United Nations Office for the Coordination of Humanitarian Affairs (UNOCHA) and aims to help develop VOST for multiple French-speaking locations during disasters, specifically VOST methodology, tools and human resources. VISOV has been active for several floods in France and was in operation after the terror in Paris and Nice. VISOV's teams have specialized their use of WhatsApp as a collaboration tool. Using color codes and 'rooms' for training, some of VISOV's chat rooms in WhatsApp have more than 50 members. For activation and alert purposes, the group uses an app called Vialert (by Cedralis), which enables mobile calls even when the device is switched off or put on silent mode.

IT'S HARD TO LIE ...

With social media, anyone in the world can watch what you are saying and thus check the accuracy of your messages. That fact might have been overlooked by the communications staff of

Finnair in the summer of July 2014, when the airline got caught for stating on Twitter that their aircraft did not fly over Ukraine. What transpired was an example of individuals and company staff monitoring what was being said on social media – evidence of the fact that there is a lot of interest in major brands such as Finnair and Flightradar24:

1. Message from Finnair on Twitter:

    'For those of you wondering, Finnair does not fly over Ukraine. Your safety is our top priority.'

2. A private person posted a message on Twitter that included a screen clipping of Flightradar24 showing Finnair flying over Ukraine:

    '@Finnair This flightradar info says differently.'

3. At this stage, of course, Finnair could have checked the accuracy of its first message and consequently issued an apology. Instead, it confronted Flightradar 24 with this Twitter message:

    'Flightradar is inaccurate and it does not take into account that the world is round.:) The route goes on the West border.'

4. Naturally, the people at Flightradar24 did not appreciate this accusation, which brought them into the conversation with this message on Twitter:

    'Flightradar24 data is very accurate. A flight over Ukraine is over Ukraine no matter if the world is flat or round.'

5. After this, Finnair came back and did issue an apology:

    'We were wrong. Apologies!'

The airline received several comments on Twitter afterwards. Some were critical of how the company had handled the situation, but one lesson learned is that the public appreciate when a company is willing to admit a mistake. Some of the comments on Twitter included:

> *'And that's how you recognize a good company – able to admit its mistake and apologize. Thumb up for you, Finnair.'*
>
> *'Appreciate accepting the mistake. Respect.'*
>
> *'I have travelled nearly 1 million miles with you. It's good to know that you also know where you are.'*
>
> *'Never let ill-informed comm people handle technical issues.'*
>
> *'Kudos for being honest. We appreciate it.'*

A good summary of the Twitter conversations – and a major learning point regarding social media – came in the form of a comment on Twitter by a private citizen:

> *'It's hard to lie in this world full of fine technology open for the public, isn't it.'*

BIBLIOGRAPHY

Andrews, C. (2016). *University of Washington, Seattle.* Retrieved from http://faculty.washington.edu/kstarbi/CSCW2016_Tweetdashians_Camera_Ready_final.pdf

Bangor Maine Police Department. (2016, December 29). *Bangor Maine Police Department.* Retrieved from www.facebook.com/bangormainepolice/posts/10154888826151079:0

CNN. (2010, January 14). *edition.cnn.com.* Retrieved from http://edition.cnn.com/2010/TECH/01/13/haiti.social.media/index.html

Cohen, S. E. (2013, March 7). *Emergency Management.* Retrieved from www.govtech.com/em/disaster/Sandy-Social-Media-Use-in-Disasters.html

Eiten, K. (2015, October 21). *KPTV.* Online efforts of virtual volunteers critical in aftermath of UCC shooting. Retrieved from www.kptv.com/story/30319594/online-efforts-of-virtual-volunteers-critical-in-aftermath-of-ucc-shooting

Gupta, A. (2013). *precog.iitd.edu.in.* Retrieved from http://precog.iiitd.edu.in/Publications_files/ecrs2013_ag_hl_pk.pdf

Hughes, A. L. (2014). *amandaleehughes*. Retrieved from www.amandaleehughes.com/HughesStDenisPalenAndersonPoliceFireSandy.pdf

Popoola, A. (2013). *irevolution*. Retrieved from https://irevolution.files.wordpress.com/2011/07/verily-swdmwww13.pdf

St. Denis, L. A. (2013). *cmci.colorado.edu*. Retrieved from https://cmci.colorado.edu/~palen/palen_papers/StDenisPalenAnderson-jeffco-iscram2014.pdf

Samenow, J. (2015, August 28). *Washington Post*. Retrieved from www.washingtonpost.com/news/capital-weather-gang/wp/2015/08/28/why-social-media-wouldve-saved-lives-during-hurricane-katrina/?utm_term=.a8494ef8f55f

Vroegop, R. (2014). *Forewarned and Forearmed: The Calgary Emergency Management Agency and the 2013 Flood*. Ottawa: The Conference Board of Canada

CHAPTER 7

Internal Communication – Don't Forget Your Employees

While handling the media is a given during a crisis, informing employees is often forgotten or neglected. Keeping staff informed, however, will empower them and make it easier to trust management decisions.

An easy way of informing employees is to send emails to everyone in the company. Consistent messaging through email does not need to take up much time, but it will definitely be valued. The message should include information about the situation, how it is being handled internally, and links to more information.

To issue more important or timely information, consider (also) using text messages to employees. Ideally you do not want to spend a long time writing down a message on the small screen on your phone, so a dedicated service for group SMS will come in handy. If not, use a PC to type the message you want to distribute, send the text to yourself by email, and open and copy it on your phone for easy input as a text message. You also need to think of ways of accessing employees' mobile phone numbers; again, a group text service or an app for sending messages to many simultaneously will be of help.

Electronic forms of communication – such as intranets or email services – might not be available in a crisis situation. You will then need to be creative and to find alternative means of communicating internally. Setting up a Facebook group or even using a special 'hidden' page on the company website are alternatives to consider, but paper-based tools such as posters or fact sheets distributed to employees in the cafeteria can also work.

FIGURE 7.1 All-employee meetings can be an effective way of disseminating information during a crisis.
Source: Harald Pettersen, Statoil.

For extreme situations, or when there is a need for employees to meet in person, arrange for a company-wide meeting. Present news and facts about the situation, and also spend some time explaining what the company or organization is doing to handle it. An all-employee meeting is an excellent venue for praising staff and for letting employees ask questions or voice their concerns.

A crisis situation is a perfect time to repeat internal guidelines for activity on social media. Remind employees that there will be rumors and fake news, and that business-related matters should not be discussed on private social media accounts. You should also use the opportunity to ask for help and invite employees to forward messages on social media that the company should react to or know about to the communications staff.

If possible, give information to employees before issuing a press release or appearing live on a newscast. This is especially important if the news concerns closing down an office building or gives information about missing or hurt personnel. You can also use intranets or group email to urge employees to watch a certain TV station or follow a press conference.

Keeping employees informed and up to date from sources *within* the organization will help them communicate correct facts about the crisis to their personal networks. As a result, you will enhance trust in leadership, have employees who value and understand your efforts, and maintain your brand in a difficult situation.

CHAPTER 8

Top-Level Communication and Management Priorities

The head of a company or organization is literally put to the test if a crisis occurs that involves employees, customers, the media or the public. How he or she handles it will have a profound effect on the outcome of the crisis, and top management communication therefore needs to be addressed before a crisis occurs.

Fortunately, there is lots to learn from previous disasters:

- Respond quickly and appropriately.
- Take responsibility.
- Go into reaction mode.
- Keep your perspectives.
- Be visible.
- Avoid speculation.
- Focus on the situation and the audience, not yourself.
- Empower your employees.
- Make sure victims are prioritized.
- Commit to communicate.

Preparing a CEO for crisis management includes discussions about possible negative outcomes that may result from the crisis, for example, the fact that the CEO personally, as well as the company or organization, will be criticized and scrutinized. There might also be campaigns attacking the management, the company or its employees – by targeting company mailboxes or through campaigns on social media.

It is vital that the CEO of a company or the head of an organization in trouble does not focus on his or her own situation. As Bill Salvin, president and founder of the consulting company Signalbridge explains, 'The audience doesn't care how inconvenient it is for *you* to work on a cell phone, they want the head guy on the ground so he can move heaven and earth to help them in their time of need' (Salvin, 2013).

THE CEO ON SOCIAL MEDIA

Top managers with a profile on Twitter or a personal blog can be an effective voice in crisis communication and a proof of concrete and visual leadership. There are several examples of CEOs tweeting with compassion after a tragic event, such as the head of Air Asia, Tony Fernandes, who posted this message on Twitter shortly after one of their flights had crashed: 'I am touched by the massive show of support especially from my fellow airlines. This is my worst nightmare. But there is no stopping.' Other leaders might choose other platforms. Prime Minister Jens Stoltenberg of Norway posted several personal messages on his Facebook page in the days following the terror in Oslo and on the island of Utoya, and he received much praise for the way he shared his thoughts and feelings.

Video can be an effective form of communicating CEO involvement, and the International Air Transport Association (IATA) therefore advises an airline to

> consider posting a broadcast-quality video statement by the CEO on its website and/or YouTube, in which he/she confirms key facts and outlines the airline's immediate priorities, with appropriate messaging. This would ensure that the CEO is 'visible' early in the response, without necessarily exposing him/her to media questions at a point where confirmed information is scarce.
>
> (IATA, 2016)

A visible leader can also assist in reinforcing important messages. CEOs and mayors often have a large pool of followers on Twitter and Facebook, and when they retweet or share a post, it reaches new and important audiences.

FIGURE 8.1 With 370,000 followers on Twitter as of July 2017, the mayor of Calgary, Naheed Nenshi, has the potential of reaching a vast audience. He actively amplified messages from @cityofcalgary and @CalgaryPolice during the flood in the summer of 2013. *Source*: City of Calgary.

THE CEO AS A SPOKESPERSON

In addition to possibly having a voice on social media, the head of a company or organization, almost without exception, needs to give a statement about a crisis on TV, on radio or in the press. To do this in a professional manner, it is vital for him or her to have gone through some type of media training. If not, tension and nervousness might hinder important messages and shine a bad light on the crisis management process. There is no doubt that rehearsing and exercises are 'insurance' for personal reputation, and media training is an investment that will always be beneficial.

In addition to following the advice on handling the media earlier in this chapter, the CEO needs to think thoroughly about which words to use to describe the situation and what is being done to mitigate it. Rehearsing and dry-runs are key preparations for an interview to be successful, and senior communications staff should advise the CEO of possible angles and questions.

A crisis situation is not the time to be funny or to play – knowingly or not – with words describing the scene. That was not the case when the head of the company responsible for a train accident in Canada (described in Chapter 1) was asked what had happened. Even though everyone knew that the cause was that 72 tank cars carrying crude oil had *blown up*, his response was: 'I think we blew it ... we blew it big time' (CBC News, 2013).

THE DIFFICULT TASK OF APOLOGIZING

> "Apology is the atomic energy of empathy. Apologies make bad things start to stop happening.
>
> James E. Lukaszewski, President, The Lukaszewski Group (personal communication)

One of the most challenging tasks a CEO will face when handling a crisis is the art of giving a genuine apology. Sometimes because they do not want to, and occasionally because they are forced to, there are many examples of companies and top managers fumbling with what should be a relatively easy task – and one that is expected. It also has a financial and reputational side, and according to an article by Michael Shmarak about corporate America, a genuine and immediate apology 'can save them billions in business and heaps of embarrassing PR blowback' (Shmarak, 2017).

As we have seen earlier, how words are chosen also plays an important role in how an apology is perceived. Some CEOs give statements like 'I am sorry if I offended anyone' or 'I am sorry if you feel that way.' That does not work. Shmarak therefore recommends that a genuine apology must include responsibility, sympathy and accountability, and that a full statement of apology must include the following six elements:

1. Acknowledge what you did wrong.
2. Take responsibility for your actions.

3. Acknowledge the impact your actions had on others.
4. Apologize for having caused pain or done damage.
5. State your future intentions and repair the damage.
6. Above all else, do not make excuses.

STAFF CARE

An important task for top management is to take care of the staff who are handling the crisis. This is easier said than done, as lengthy meetings, business continuity and financial questions often occupy many hours for a leader dealing with a crisis. However, visible leadership also *within* an organization is valued and should be prioritized. It does not take much, and a surprise visit to the communications team to cheer them on and give encouragement goes a long way in providing support. Handing out bags of candy or making sure they get pizza when hungry are other ways for the top manager to show his or her commitment. Praising staff in public and in media interviews is also a winning gesture.

More serious is the fact that communications teams and other internal staff handling a crisis might need psychological follow-up after the crisis has subsided. Traditional first responders usually get a chance to see a psychologist or have a debrief, but those who respond digitally should also be considered. At the same time, looking after personnel in this way will build resilience and make the organization even better prepared for another crisis.

CHOOSING YOUR WORDS

Crafting a quote for a press release or deciding key messages for an interview, a speech or a funeral takes time, knowledge and a mastery of words. If possible, senior communications advisors should write and rewrite, adjusting words and sentences until they bring about exactly the feeling the victims, the journalists or the public need to hear.

There are many excellent speeches given by CEOs from around the world, but I choose to end this chapter with the last sentences from Norwegian Prime Minister Jens Stoltenberg's introduction to the press conference on the evening of July 22, 2011:

We must never give up our values.
We must show that our open society can pass this test too.
That the answer to violence is even more democracy.
Even more humanity.
But never naivety.
We owe this to the victims and their families.

BIBLIOGRAPHY

CBC News. (2013, July 9). *CBC.CA*. Railway chair defends safety record. Retrieved from www.cbc.ca/player/play/2396202194

IATA. (2016, December). *iata.org*. Retrieved from www.iata.org/publications/Documents/crisis-communications-guidelines.pdf

Salvin, B. (2013, July 17). *Signalbridge*. Retrieved from http://signalbridge.blogspot.no/2013/07/tragedy-at-lac-megantic-rail-chiefs.html

Shmarak, M. (2017, May 9). *ragan.com*. Retrieved from www.ragan.com/Tomorrow/Articles/For_a_genuine_apology_go_back_to_the_basics_52725.aspx

CHAPTER 9

High-Flying Crisis Communication (the Special Case of Airlines)

The airline industry stands out in the crisis communication world for several reasons: not only because it has seen its share of tragic events, but also because there have been several instances of 'live reporting' from passengers on airplanes *while* they are experiencing an emergency. With wi-fi on board, passengers are able to surf the web or update their Facebook page 30,000 feet up in the air, and the same technology can, of course, also be used for tweeting messages about the plane having problems or live-streaming an emergency landing. This kind of broadcasting obviously brings new challenges, especially for the airline's communications team, and it certainly makes for several lessons learned and best practices that can be useful to consider.

T+15

With 275 members from 117 nations, the International Air Transport Association (IATA) offers guidelines and advice for inter-airline cooperation, some of which focuses on crisis communication. In 2012, the organization published its first guidelines on social media, and the document has since been restructured to include new technologies and updated advice based on recent events in the industry. In the guidelines' introduction, the organization explains some of the challenges an airline faces after an accident: 'The airline may be left struggling to make its message heard above the cacophony created by citizen journalists, politicians, government agencies, celebrities, "experts" and self-publicists eager to share their opinions' (IATA, 2016).

The IATA document has a chapter on 'Communication flow and timeline,' which describes how an airline should strive to communicate after first being notified (T = 0):

T+15 mins: Release first 'tweet' acknowledging initial reports. Update regularly with short posts as new information is confirmed.

T+ 60 mins: Issue longer summary of information confirmed to date, via multiple channels and posted on website. Release new summaries hourly, or as key developments are confirmed, while maintaining regular flow of short updates.

The concept of 'T+15' is one to consider also for crisis managers in other industries. Although very few will be able to fulfill its goal, it says something about the need for being prepared and the importance of social media surveillance 24 hours a day, seven days a week …

CHALLENGES TO CONSIDER

It is not only the advent of social media that has brought challenges for an airline's crisis management team. Several flight tracker websites make it possible for anyone to follow the route of an aircraft – in real time – and to look up its history and last known position and speed. It is also possible to subscribe to flight emergency messages on Twitter, and several airline operation rooms have a dedicated screen displaying such messages.

Flightradar24

Flightradar24 started as a hobby project in 2006 when two Swedish aviation geeks decided to build a network of ADS-B receivers in Northern and Central Europe. It is now the largest such network in the world, with over 10,000 connected receivers. Flightradar24 tracks over 150,000 flights a day and has more than 1.5 million daily users.

An airline's communications team needs to be aware of several special challenges concerning the industry:

- Passengers who are on board a flight having trouble might call the airline or family members before any of the airline staff know something is wrong. Distressed passengers can also get in touch through text messages or via social media.
- Be aware of 'accidental spokespersons' – such as passengers tweeting about an emergency – becoming important sources for journalists.
- Other airlines might express their sympathy before your own has issued a statement.
- Changing the top banner on social media accounts should be considered.
- Flightradar24 and other flight tracking services are important sources of crisis information – for the airline as well as for passengers, media and the general public.
- Other parties involved – such as airports or the National Transportation and Safety Board (NTSB) – will probably start commenting early, and their social media accounts should therefore be followed closely.
- Families of passengers on board a flight that has had an accident are likely to show up at the flight's departure or arrival airport. Airlines therefore need to be prepared for arranging two Family Assistance Centers – one at each airport.
- Arrangements for a Family Assistance Center should be handled and communicated promptly. (The airline will have to 'compete' for suitable hotels with government agencies and media.)
- Use social media for customer service.

THE ROLE OF NTSB

The U.S. NTSB plays an important role in aviation accidents in the U.S., or when a U.S. carrier has an accident abroad. Their primary role is to investigate, but NTSB will also actively issue information about their work. Airlines that are involved in an accident that NTSB will investigate should therefore work closely with the agency and be acquainted with their routines for disseminating information.

The NTSB is active on social media and will often post on Twitter within the first hour of an accident. Usually, this first message is a confirmation that something has happened, and NTSB might also publish a statement about their Go-Team having been sent to the crash site. Later, the NTSB will also arrange press conferences, give out factual bites of information and provide photos of on-scene work. NTSB will often live-tweet and provide video from their media briefings.

In addition to investigating and communicating about an accident, the NTSB has its say in taking care of next-of-kin, in close cooperation with the airline. The NTSB manages the Federal Family Assistance Plan for Aviation Disasters, which clearly points out who is responsible for family assistance:

- The air carrier has a fundamental responsibility to victims and their families affected by an aviation disaster. The air carrier is primarily responsible for family notification and all aspects of victim and family logistical support.
- All personnel involved in providing services to assist victims and their family members should be trained in crisis response and must demonstrate compassion, technical expertise, and professionalism.

LAWS AND REGULATIONS

Today, there are strict laws in the U.S. and the EU regarding how an airline must provide care for next-of-kin after an aviation disaster. The Disaster Family Assistance Act of 1996 and the Foreign Air Carrier Family Support Act of 1997 are important documents that came into being as a result of hard work and dedication by several individuals who did not receive adequate treatment after having lost family members in airline accidents. Paramount in this development was the accident with Flight 4184 in 1994. According to the support group that was formed after the tragedy, there were several examples of mistreatment and unprofessionalism (American Eagle Flight 4184, n.d.):

> Following the crash of American Eagle Flight 4184, several areas of critical importance to the families were mishandled. The victim

> identification process fell short. Although 68 caskets containing small amounts of the remains of each victim were distributed to all families, for most, their loved ones were buried in seventeen caskets of 'unidentified remains.' These seventeen caskets were interred at night, without seeking the consent of the families nor informing them of the interment. Additionally, at least one family received remains that were misidentified. Families learned that most of the identifiable personal effects of their loved ones were destroyed rather than being returned to them. The cleanup efforts at the crash site lacked coordination and were incomplete. Families and friends who visited the site months after the clean-up was completed retrieved human remains, personal effects, and plane parts.

After much pressure from support groups and individuals, President Bill Clinton in 1996 signed the Disaster Family Assistance Act describing family assistance in detail. The next year, The Foreign Air Carrier Family Support Act became law, requiring foreign air carriers to transmit to the Secretary of Transportation and the Chairman of the NTSB a plan 'for addressing the needs of the families of passengers involved in an aircraft accident that involves an aircraft under the control of the foreign air carrier and results in a major loss of life' (Congress, 1997). The plan submitted by the foreign air carrier must include the following:

- A plan for publicizing a reliable toll-free number for the passengers' families
- A process for notifying, in person if practicable, the families of passengers involved in an aircraft accident
- An assurance that such notice will be provided to the families as soon as possible after the identity of a passenger has been verified
- An assurance that such carrier will provide and update a list of the best available information regarding the names of the passengers aboard the aircraft
- Assurance that the family of each passenger will be consulted about the disposition of any remains and personal effects of the passenger
- If requested, return to the family of any possessions of the passenger

- Assistance to the family to ensure that they are able to travel to and stay at the location

The United Nations also focuses on family support, and in 2013 its organization ICAO (International Civil Aviation Organization) published 'Policy on Assistance to Aircraft Accident Victims and their Families.'

WHAT TO SAY

With so many laws, regulations and 'players' involved, communicating about an aviation disaster requires cooperation and a clear understanding about who says what. There are strict rules about who can release investigative information, and instead of getting involved in detailed discussions about cause and effect (which can also have an impact on law suits and compensation requests), the airline should therefore focus on other topics:

- Condolences
- Arrangements for survivors
- Care for employees
- Actions for customers not involved, for example rerouting
- Cooperation with authorities such as NTSB
- Effect on operations
- Actions taken
- Information about safety protocols and processes

Airport Cooperative Research Program

In 2017, the National Academies Press published a research report titled 'Establishing a Coordinated Local Family Assistance Program for Airports.' The report's 144 pages cover a variety of advice for airports regarding how to assist families after an airline accident and include sections such as 'Roles and responsibilities,' 'Information management,' 'Communicating with affected families' and 'Recovery.' In addition to the report, the authors have published check lists and training material on their website.

SOCIAL MEDIA IN AVIATION

Not having routines in place for monitoring social media on a 24/7 basis can have a negative impact on reputation, management and stock prices. One example is an engine failure on board a Qantas flight in 2010, when – according to a report by SimpliFlying – speculations ran amok on social media as a result of no authoritative voice from the airline:

> On 4 November 2010, QF32 had an engine failure after taking off. Qantas senior management, however, became aware of this only when its share price 'started to collapse' as a result of rumors of a plane crash on Twitter. These were all happening while the aircraft was still flying back to safety.
>
> (Guen, 2015)

A more serious event took place in January 2014, when a flight to Singapore experienced technical problems and started an emergency descent. With gas masks hanging from the ceiling and frightened passengers everywhere, several people on board took pictures of each other and posted them on social media. One of them was a U.K. citizen named Nathan Phelps, who took a picture of other frightened passengers and then also took a selfie, which became an immediate 'hit' on social media. Many people commented that the situation could not have been that bad if passengers had time to take pictures of each other. Phelps explained later:

> The truth is that when that photo was taken we knew that we were going to be OK, the pilot had told us the issue and was just looking to drop a further 5000 feet and also find a suitable airport at which to land at.

The public treated it like some kind of joke, but the situation was far from funny. Phelps said:

> If they had seen the other photo I took when the plane was shaking as it descended unnaturally, they would have seen the terror on the passengers faces. That photo was also on my time line but people only chose to comment on the selfie.

FIGURE 9.1 With social media connected to wi-fi on board commercial flights, passengers can broadcast live from emergency situations – or take a selfie.
Source: Nathan Phelps.

He later explained his reasoning:

> I have questioned myself many times afterwards about what drove me to take a photo when I genuinely thought I was going to die. The only answer I can come up with was that I thought there may be a chance that someone may recover my phone and see what we went through just before we crashed ... Fortunately, it never happened.

One airline that has taken 'online listening' to a high level is Southwest Airlines, which has a state-of-the-art 'listening center'

FIGURE 9.2 Southwest Airlines has established its own 'listening center.'
Source: Southwest Airlines.

where staff monitor conversations 24 hours a day, seven days a week. Their goal is to listen and engage in conversation around the brand and industry, allowing Southwest to be 'ultra-responsive.' The center monitors news coverage, social media conversation and employee response on the airline's internal intranet website.

RESILIENCE IN AVIATION

Few people in the aviation industry have had more experience in handling tragedies than Ken Jenkins. For many years he was head of the American Airlines CARE Team, which consists of volunteer employees willing to respond to airline accidents. In ten years, Jenkins responded to eight fatal events, giving him a unique insight into next-of-kin support and crisis management. Jenkins has described his experience from the disasters in his book *Resilience*, and he has also published several podcast episodes called 'The Black Box.'

FIGURE 9.3 Ken Jenkins handled eight fatal events for American Airlines.
Source: Kjell Brataas.

BIBLIOGRAPHY

American Eagle Flight 4184. (n.d.). Aviation Disaster Family Assistance Act of 1996. Retrieved from www.americaneagle-flight4184.com/adfaa-of-1996.html

Congress. (1997, September 18). Retrieved from www.congress.gov/105/bills/s1196/BILLS-105s1196is.pdf

Guen, L. (2015, May 19). *SimpliFlying*. Retrieved from http://simpliflying.com/2015/airline-crisis-communications-accidents-incidents

IATA. (2016, 12). *IATA.ORG*. Retrieved from www.iata.org/publications/Documents/crisis-communications-guidelines.pdf

CHAPTER 10

Family Support and Victim Assistance

INTRODUCTION

Supporting families and victims after a tragic event is not only crucially important, but also one of the most challenging tasks a responsible company, organization, government agency or ministry can face; not only because the need for information is overwhelming, but also because satisfactory answers might be lacking or worse, involve news that will devastate those receiving it. A contributing factor is that communications specialists themselves will be personally involved, even touched, a fact that is natural and unavoidable, but still puts an extra constraint on the communication efforts.

Victims can comprise anything from one person to several thousand, first graders or a group of retired people. Nevertheless, there are certain notions that relate to all, as described in a pamphlet from Victim Support Europe, which recommends that every victim (Victim Support Europe, 2015)

- Is treated with respect and dignity
- Is able to access the information they need
- Is able to access support services
- Is able to make their voice heard throughout the criminal justice process
- Has access to strong rights
- Receives the compensation they deserve

Victims are sometimes portrayed as helpless and destitute, but even if they are in a desperate situation, victims can be aided to cope and deal with the situation. To achieve this, it is vital to focus on *resilience*, and to tell victims that they will be taken care of and that someone is looking after them.

There are many reasons for treating victims right, as victim assistance can sway issues such as

- The crisis itself (and – obviously – the victims themselves)
- Public perception
- Media coverage
- Trust in management
- Local communities
- Recruitment
- Affiliate organizations
- Stock prices
- Legal issues
- Compensation schemes

It is a fact that unhappy victims often receive attention from the media, and with social media, victims now also have a direct, unfiltered voice that can get worldwide attention. However, the most important reason for taking care of families and victims is that it is *the right thing to do*.

Communications staff are often asked to advise top management and HR personnel about what to say, how to say it and in general, how to inform victims in a professional and compassionate way. Luckily, most of us have never been in a situation that requires this type of communication skill, which makes it even more important to be prepared. This chapter will therefore give a thorough analysis and best practices for communicating issues dealing with life, death and uncertainty.

PREPAREDNESS

The first advice is to be as prepared as possible. This involves a lot of 'what if' scenarios and a focus on worst case. Recent events – and many of the case studies in this book – show that a crisis can happen, and that you have an advantage if you have also planned for the human side of a crisis. It will never be routine, and not everything can be planned or anticipated, but it will be so much calmer if you have prepared for the unexpected, including the support of victims.

Communicating with family members and victims is often challenging. One reason is that the receiver of the information you are giving might be a traumatized victim feeling emotionally overwhelmed. Often, they do not 'take in' facts given, meaning that you should repeat messages and distribute them in an oral as well as a written form.

You also need to prepare for confusion, especially in the early stages of a tragedy. Facts might be hard to find and confirm, and in many instances those affected by a disaster have had their lives changed – forever. Their thoughts are therefore a mixture of hope and despair, and at the same time they are bombarded with news flashes and rumors. As time progresses, they also have to handle totally new concepts and decisions, such as talking to a coroner, dealing with a claims form, discussing autopsies with the police and getting notifications of remains. As a 9/11 widow said, 'They gave me all these forms to fill out, but I couldn't even remember my name.'

While confusion is hard to avoid, being prepared to disseminate information regularly and as soon as news breaks might alleviate the situation and in some ways fill the void victims experience regarding what is happening and the next steps. In the book *Collective Conviction*, members of a British organization called Disaster Action state that the need for information 'is one of the most fundamental, urgent and significant needs of those involved in disasters' (Dix, 2014). The authors explain that in the immediate and short-term aftermath, information should cover who, what, where, when and how the disaster happened. They conclude that 'In the longer term, information needs tend to focus on finding out why the disaster happened, whether it might happen again and learning lessons.'

> "
> The expectations of accident survivors and families of victims is steadily increasing as a result of the activities of family advocates and associations (particularly those formed after previous accidents) and plaintiff's attorneys.
>
> IATA guidelines (IATA, 2016)

VICTIMS FIRST

Although not always attainable, a good practice is to always inform victims before you give the same facts to the media. How this is done should be thought of in advance, of course, and could involve regular presentations at a Family Assistance Center or in some cases messages posted in a closed Facebook group.

The Global Counterterrorism Forum suggests following certain guidelines for making sure that information is handled correctly (Global Counterterrorism Forum, 2012). They include

- *Immediacy* – intervening as soon as possible
- *Accessibility* – making assistance convenient and available
- *Simplicity* – using quick and simple methods adapted to the situation
- *Unity* – identifying an official point of contact for the victims
- *Resiliency* – helping with victims' self-esteem and their coping strategies
- *Comprehensive assistance* – taking into consideration all the particular needs of the victims

In preparing for victim assistance, a useful approach is to foresee questions from next-of-kin and survivors, and to prepare for answering them. Such questions can include:

- What will medical care be like?
- Who will pay?
- What does my insurance cover?
- Will there be an autopsy?

- Where is the body now?
- What legal actions do I need to take?
- How does compensation work?
- When and how can I retrieve personal effects?

Knowing what is important to victims also helps plan for the right type of assistance. Often, next-of-kin want to see where their loved one died, and getting back personal items of the deceased is also high on their agenda. This may not seem important to others, but for a family member it can be crucial to salvage the last thing their family member touched before he or she died.

FBI's Office of Victim Assistance

The terror on 9/11, 2001, prompted the FBI to review the way it handled victim assistance, and since then the bureau has had a robust program for victims of a variety of crimes. The Office of Victim Assistance has about 150 victim specialists in field offices throughout the U.S. who respond to a variety of crimes, including cybercrime, domestic abuse, murders and mass fatalities. The specialists are part of an investigation team, but they also provide information and support to victims. They are specially trained to respond to mass casualty events, including death notification, victim lists and the return of personal effects.

INTERNAL PREPARATIONS

Internally, organizations can prepare in a variety of ways. Depending on the size of the company or government organization and its operations (local, national, international etc.), some or all of these questions should be answered in the planning phase, and before something happens:

- Is our next-of-kin database updated and accessible?
- Who are our internal volunteers for victim assistance?
- Which languages are spoken by employees' next-of-kin?
- Who will pay for travel?
- Who is in our Go-Team?

- What costs will be incurred?
- Whom do we interact with?
- Who can help us?
- How do we take care of our staff?

The FBI's categories are one way of dividing victims into groups:

- Missing
- Deceased
- Injured
- Present but not injured
- Business community

VICTIM ACCOUNTING

The question 'Who is a victim?' is not so easy to answer, as we no longer count only the 'walking wounded' in this category. The wounded and the family members of those who died are easy to identify, but what about bystanders – who are often first responders – and witnesses who saw a tragic event unfolding? Adding to the confusion is the fact that many individuals do not self-identify as victims or feel they do not deserve attention. In New Zealand, this was handled by using social media to actively ask victims of the earthquake in 2014 to come forward and to register with authorities. On the opposite side of the scale are the few instances when victims come forward and they are told the government does not consider them victims, a situation that can mean retraumatizing them, as the person feels excluded from the rest of the group.

Knowing the status and location of all victims is a challenging task, especially when a crisis involves a large number of people. One example is the Asiana plane crash in 2013, which involved 214 patient transports in 52 ambulances. Eleven different hospitals received a total of 198 patients, and it took approximately ten days to determine with certainty where each patient had been admitted (NTSB, n.d.).

There is no doubt that the most important of all victims' lists is the one that confirms who died. But chaos, patients at several hospitals and inadequate list management often contribute to an unclear overview of who has died. This was the case in several countries in Scandinavia after the tsunami in Asia in 2004, when authorities used a mix of spreadsheets and notes on paper to systemize information about casualties. 'Spreadsheets is a four-letter-word when it comes to victim accounting,' says Mary Fetchet, who is the founding director of the support group Voices of 9/11. She recommends using a more sophisticated database, and fortunately, several government agencies and private companies in the field of crisis management offer web-based systems for storing and accessing data about victims and their status.

Victims in Nice

Identifying victims can be challenging, as was the case after the terror in Nice in 2015. Some 30,000 people were on the open space of the promenade when a truck killed 86 people and wounded several hundred, but who were the victims? The Paris prosecutor, responsible for establishing the list of victims, considered that people had to prove they were on the truck trajectory to be considered victims – they had to have been targeted by the assassination attempt. More than 2,000 letters were sent out to potential victims, who had to somehow prove (by providing copies of SMS messages or pictures taken at the time) that they were in the area when the attack took place. By December 2016, 94% of the injured victims and 88% of the legal successors had received a provision. The process was helped by the fact that that France has instituted an Inter-Ministerial Cell for Victim Support (CIAV), in charge of victims' information and coordination, which includes a single database about casualties.

VICTIMS ABROAD

When an individual or a group of people die or become victims abroad, several additional challenges become apparent. Time zones and cultural differences must be considered, embassies or consulates need to be involved, and victims' families will often have a hard time finding information and getting in touch with their loved ones. Language is another important barrier, as it can be hard to get in touch with local hospitals or law enforcement agencies when the staff there might not speak any foreign languages.

At the same time, those who have experienced a tragic event in a foreign country might not have access to their usual support network. Care and practical help from the employer, the government or victim assistance organizations are therefore necessary – and often expected.

With a multitude of languages and nationalities among the victims, communicating after terror attacks becomes especially challenging. To prepare, several governments and organizations have produced documentation about what foreigners should do if they become victims abroad. One example is 'Where to get help if becoming victim of a terror attack in Germany?', issued by the German nationwide support group Weisser Ring.

Another consideration is the treatment of the dead in a foreign country. While most countries in the Western world have adequate facilities for storing bodies, that may not be the case in other parts of the world. Identification processes also vary widely, and in many countries pictures of dead bodies are displayed publicly so that relatives may be able to identify their deceased family members. That was the case in Thailand after the tsunami in 2004, when hundreds of pictures were on display in temples throughout Phuket and Khao Lak.

TELEPHONE HOTLINE

Soon after an accident or other tragic event, a good practice is to set up a telephone support system that relatives of those who might have been involved can call to get information or report their loved ones as missing. Organizing such a service is not an easy task, and planning for a telephone hotline must therefore be done *before* something happens.

A few of the questions relevant to organizing a hotline are:

- What technical system do we need for handling a large number of incoming calls?

- Who will be answering?
- Where will the phones be answered?
- How do we save and process messages and information from the callers?
- What languages can we expect callers to speak?
- How do we take care of our staff who answer the calls? (Breaks, food, follow-up etc.)

It might sound trivial, but in the heat of the moment it is of the utmost importance and help that the staff answering have basic tools, such as a headset each, so that they can focus on the caller and register what he or she says in a predefined form that lists the basic information needed for further processing and assistance. There could also be some kind of 'back office' that handles follow-up and database registration of the information coming in. If there are enough staff and funding, a good practice is to have *two people* answering each call – one who focuses on the conversation, while the other registers the essence of what is being said.

No matter who or how many respond, training is needed *before* they start receiving calls. As a minimum, each hotline operator should receive a few hours of training and discussions about what to expect and how to effectively communicate with the callers – who might be desperate, in disbelief or crying. Training should therefore focus on how to respond and what to say.

They should learn that it is OK to give out basic facts about what has happened and what is being done over the phone, but it is vital that the hotline staff do not engage in thoughts about worst-case scenarios or discussions on who is to blame for what has happened.

Death notifications should never be carried out over the phone – and usually involve the police and a minister. Staff should nevertheless get some basic training in being the first to give a family member news about an accident or that someone is missing, as next-of-kin always prefer to get bad news from someone 'official' instead of learning about it on the TV news or as a message on Facebook. If there has been a fire at the office, or if someone from the company's staff is missing after an accident abroad, company representatives should be the ones informing next-of-kin. To minimize the number of calls coming in, it is of great help if those who are safe or not seriously wounded can call their relatives *themselves* without involving the hotline staff. Such a 'phone home' policy could be part of internal manuals and training, so that employees react appropriately. If available, other digital tools should be used, for example Facebook Safety Check or Google Person Finder.

When talking to those who connect through a hotline, the staff should concentrate on being respectful and accommodating. It is important to register concrete needs from the person calling, and to get as many facts as possible about the next-of-kin they are calling about. One example could be facts about where and how they were last in contact or at which hotel he or she was staying.

Depending on the nature of the event in question, it is usually never advisable to promise to call back. Demand at the hotline can quickly escalate, meaning that it can be hard to find time to make outgoing calls. Instead, next-of-kin should be directed to a Family Assistance Center, or a closed web page or Facebook group where more information and news can be released regularly.

Answering incoming calls at a telephone hotline is psychologically challenging and can be physically draining. Staff should therefore get a break after one to two hours. After six to eight hours, they should be relieved for the day. As mentioned in Chapter 2, I have personal experience of staffing a telephone hotline after the tsunami in Asia. This was poorly organized and executed, and we learned the hard way about the importance of planning in advance, having adequate database systems, and simply feeding your staff and giving them enough breaks during each shift.

FAMILY ASSISTANCE CENTERS (FACS)

If a crisis has the potential for many hurt, missing or dead victims, a Family Assistance Center should be considered. Such a center has many names; in the U.K. it is often referred to as a Humanitarian Assistance Centre, and in the U.S. it is sometimes called a Community Resilience Center. No matter what we call it (for the rest of this chapter we will use the acronym FAC), such a center is a gathering place for those who seek information and comfort after a tragic event. It is often found in a hotel close to where something has happened, but an FAC can also be organized in part of an airport or at a corporate head office.

As a one-stop shop for information, an FAC can have many advantages for victims as well as organizers:

- Information about the incident, including new facts, ongoing investigation and so on, can be given to many people at once, and before informing the media and the public.
- An FAC provides a safe, private space to meet with others affected by the same event.
- Family members can be given facts and advice about what will happen next.
- Emotional and practical support can be offered.
- Families can draw support from being with others in similar circumstances.
- Families can receive advice on legal matters, insurance, compensation and so on.

As with setting up a telephone support system, making an FAC functional requires planning and a lot of work in advance. A few items to consider are:

- Location – and alternate location (often hotels)
- Staff and shift planning
- Security and check-in procedures
- Printed signs and arrows
- Circular tables (more inviting to conversation than square ones)
- Communication at the center
- Visits by CEO/management/other leaders in charge of operation
- VIP visits (prime minister, president etc.)
- FAC management
- Handling media at the center
- Rules for social media, pictures, video etc.
- Taking care of staff
- Name tags, (blue) vests, cordons etc.
- Invited helpers, e.g. ministers, social workers, psychologists etc.

It is important that the FAC has enough space or is flexible, as there might be a large number of people coming to the center. Often, 10 to 15 family members show up for one person dead or missing, and they come in all ages, ethnicities and religious faiths.

Management and staff at the FAC should spend a lot of time telling families at the center about the *process* they are now part of. They need to know what will happen and the order of events for the next hours and days, and being at the FAC can help navigate the process. Setting expectations early and sharing news regularly also promote confidence in the response.

Personal meetings and one-on-one talks with victims are mentally draining, and staff at the FAC should be prepared for family members being aggressive, accusatory or in denial. Emotional people often have a hard time processing information they are given, and it is therefore vital to repeat important messages and to distribute information also on paper. 'You heard the press briefing, did it make sense to you?' might be a good way of starting a conversation to check that families at the center have understood what was said. Staff at the center should be careful about making promises and should be prepared to give out information that can be hard to accept:

- News about loved ones can take hours, days or weeks to confirm.
- Return of belongings can take more than a year (especially if there is to be a criminal investigation).
- Identification of victims can take days, months or years.
- There might be massive media attention (offer assistance with handling journalists).
- There will be lots of paperwork and forms to fill out (offer assistance with this task).

There is no 'formula' for finding the right words to use when talking to a family member who has just lost a loved one. Sometimes, compassionate silence can work too. Leif Hafnor was one of the psychologists volunteering their services at the Sundvollen Hotel near Utoya, which functioned as an FAC, in the evening of July 22, 2011. In an interview with NRK TV, he explained his experience:

> It was a very chaotic situation. Young people sat in apathy by themselves with blankets draped around their shoulders. Some were comforting each other, crying. I especially remember one boy who sat by himself looking drearily out the window. I sat next to him and thought to myself that is not possible to say anything or to explain in any way that will lessen his pain. What we can do is to be there, be one they can hold on to.

Because it involves people's emotions and tensions run high, an FAC needs a large number of staff. At the center in Marseille organized after the Germanwings accident, about 90 Lufthansa and Germanwings staff worked with next-of-kin, and at the hotel in Bergen in Norway where Statoil had its FAC after the In Amenas incident, 100 Statoil employees worked for more than a week. How they are organized and what their functions are called varies greatly from incident to incident and from country to country, but a template that might be useful can be found in The Family Assistance Center Plan of Los Angeles County (Los Angeles County, 2014). It describes organization charts for a minimal level and an expanded level of service, listing a total of 39 different positions or units at an FAC. This is obviously an exhaustive list that might be too much, but it is useful to review the list of positions so that the multitude of needs at an FAC can be thought of and planned in advance. Many organizations use internal volunteers with great success, but they need training and advice before they start dealing with victims.

Being the director of an FAC is especially demanding. There is a constant need for communicating precisely and making decisions, and the director has to handle emotions from families as well as staff. Such a job should therefore be on a rotation basis. Breaks and debriefs are important for everyone working at an FAC, and to avoid staff burn-out, no one communicating directly with families should work at an FAC for more than eight hours at a time.

> "
> Early warning of stories before they break in the media, advice on contact with journalists, accurate and verified information in the face of rumor and speculation are all essential elements of next-of-kin support.
>
> Statoil's In Amenas Investigation Report

Virtual Victim Assistance

In many instances it is not possible to bring all victims' families to one physical location. It is therefore a good idea to consider setting up some kind of *virtual* center to which next-of-kin can get access.

This means that family care will be more inclusive geographically, and another benefit is that important information and news can be spread to a defined group simultaneously.

There are several ways of doing this. The quickest and easiest is probably to set up a closed Facebook group and then invite those who need it to take part in the group. Another option is a web page, which can have an 'open' area and some pages that can only be accessed by next-of-kin with a (double) password.

An example of such a website is one that the Dutch organization Slachtofferhulp Nederland organized after the shooting down of flight MH17 in 2014. With 500 paid staff and 1,100 volunteers, Slachtofferhulp plays an important role in victim assistance in the Netherlands, and it has templates and empty web pages ready in case of a tragedy. 'We call the MH17 incident 'the Dutch 9/11', and it made a huge impact on society,' says the director of Slachtofferhulp Nederland, Victor Jammers. His staff can have a new website ready at two hours' notice, and families of the passengers on MH17 were therefore able to access important information about the tragedy within a short while after it occurred.

Their website had an open and a closed section. It evolved gradually, as specific needs for facts and advice became apparent. As of this writing, the website still exists, with the following information available in the open part accessible to everyone:

- National remembrance ceremony
- Grief and loss
- Remembrance and saying goodbye
- Practical information
- Legal matters
- Compensation
- The media
- Children and youth
- Friends, relatives and colleagues
- Members-only section for next-of-kin

The part of the website that can only be accessed by victims offers more detailed information as well as a forum where members can share their stories or ask questions.

Template for FAC Hand-Out

Victims of a traumatic event are often in shock and have difficulty in processing information they are given. It is therefore important to repeat messages and to distribute information in a variety of forms. At a Family Assistance Center, organizers should have a written document to hand out or place on each table that gives important information to guests. Below is a suggested template:

Welcome to the Family Assistance Center

We hope this center will be of help for victims of the tragic events (details). We will do everything we can to make your time here as useful as possible.

We can assist with:

- Updates and news about what has happened
- Someone to talk to
- Answers to practical questions
- Referrals to experts, police, clergy etc.
- A light meal and refreshments

With blue vests, our staff are easily recognizable. Please get hold of them if you have questions or requests, but bear in mind that we often lack concrete information or facts about individuals.

Everything you tell our staff is confidential.

So that visits to the center will be OK for everyone, we ask that you comply with our basic rules for staying at the Family Assistance Center:

- Taking pictures or videos is not allowed.
- We ask that what happens here and what is said at the family assistance center is not distributed on social media.
- Families or individuals who want to give interviews to journalists must do so outside of the Family Assistance Center (our staff can provide assistance and advice).

Please do not hesitate to contact us if you have questions.

SUPPORT GROUPS – A COLLECTIVE VOICE FOR VICTIMS

Sooner or later after a tragic event, survivors and next-of-kin often feel the need to come together for a meeting or to organize a support group that can facilitate the sharing of feelings, discussions on practicalities, and helping each other to make sense of what has happened. In addition, participants will find that the group is useful for discussing available resources, insurance claims, compensation and media attention. Later on, support groups can also play a role in discussing site visits and memorials, and by being members of a group, victims can influence decisions. Every victim wants to see prevention work, and many join a support group in order to make sure that the same kind of tragedy never happens again. Meeting with government officials or private companies can therefore be an important goal of the support group.

One of the First Support Groups – the Aberfan Young Wives' Club

In 1966, coal waste slid into a local school in Aberfan, killing 28 adults and 116 children. The women of the community met up a few months after the disaster to provide mutual counseling, and their weekly tea-and-chat gatherings at the Aberfan Hotel continue to this day.

Finding prospective members of a support group can be challenging. The police are seldom willing to give out lists of victims, but traditional media and messages on Facebook are useful tools for reaching out. A government agency, such as a municipality or an embassy in a foreign country, might also be able to contact victims and inform them that a support group is being planned or offered.

When it works as planned, a support group can be a collective voice for victims, but such groups are often prone to discussions among their members and disagreements between the next-of-kin of those who died and survivors. A variety of 'group dynamics' come into play, depending on the number of members, whether a company or organization is at fault (such as after a plane crash), or because members cannot agree on handling the media or what to 'demand' from the government. The type of tragedy the group has experienced also has an impact; if victims came from the same company or school, they will have more in common than if they represent victims who do not have anything in common other than – probably by chance – being on the same plane, boat or bus.

Jelena Watkins, a psychosynthesis psychotherapist in the U.K. and a founding member of the U.K. 9/11 support group, has found that support groups often face three major challenges:

- Data protection
- Funding
- Group members' disagreements

Another possible downside is the type of information shared in meetings, as too many traumatic details might upset some members, while others want to know as many facts as possible. There might also be discussions about leadership, media attention and eventually, for how long the group should exist. However, the most difficult challenge often has to do with financial matters, as members may have different views on topics ranging from the group's funding to compensation claims from individual members.

Depending on the type of tragedy, a government agency or a private company might be heavily involved in discussions and collaborations with support groups. The topics mentioned above are worth bearing in mind when planning how to interact with representatives from a group.

To alleviate the challenges connected with support groups, it is important to know about the pitfalls in advance and to consider having an outside expert take part in some of the meetings. Discussions about media statements and focus areas should be held early on. Other steps to take include the following practical considerations:

- *Create a logo*. There might be internal talent within the group members who can design a simple logo. If not, consider arranging a competition among those affected by the tragedy, use a crowdsourcing service or ask a professional design agency for a pro bono contribution.

- *Consider an advisory board.* It always helps to have someone from 'outside' who can give advice and view the topics for discussion from several angles.
- *Develop a database.* Contact information, facts and special needs of victims can soon spiral out of hand if you try to use a paper-based system or a simple spreadsheet. A database has many advantages and can be of great help in sorting member lists and cross-referencing information.
- *Arrange fundraising events.* One of the major challenges of a support group is to find financial means to pay for meetings, office supplies, travel and so on. While membership fees are one solution, a well-organized event – preferably with some high-profile guests – can mean quite a lot of extra cash. At the same time, such an event can be a positive focus for victims and a way for them to get to know each other.
- *Communicate professionally.* How you communicate with members and the public can mean a great deal to how it is perceived by the public and its members. Use several tools, such as group emails, newsletters, the web and social media, to get your messages across.

Learning from VOICES

An organization worth looking into for learning and advice is the Voices of September 11th. It began informally in 2001 with a small group of next-of-kin meeting in the home of Mary Fetchet, who had lost her son Brad on 9/11. The meetings took place weekly for over a year, and in December 2002 the first VOICES office opened in Connecticut (Voices of September 11th, 2014).

Voices of September 11th had a clear mission: 'To serve as a clearinghouse of information, providing a wide range of support services for all those impacted by 9/11, and promoting public policy reforms to make the country safer.' To achieve their goals, the VOICES organization established several coalitions that advocated on behalf of the families, they launched teleconference support groups, and social worker staff offered confidential anxiety and depression screening over the phone.

Membership in VOICES grew to more than 16,000 individuals from all over the world.

FIGURE 10.1 Heidi Snow's organization ACCESS has 250 'grief mentors,' who provide personal support to victims after airline accidents and other events involving sudden loss.
Source: ACCESS.

Access to Peer Support

When Heidi Snow lost her fiancé, Michel, in the 1996 TWA 800 crash, her life was forever changed. In her book *Surviving sudden loss: Stories from those who have lived it*, she explains that there were no air disaster support groups, so she located a meeting of the families of Pan Am 103, which had occurred eight years earlier. She immediately felt a connection with the families of Pan Am 103, especially with a woman who had also lost her fiancé. Realizing the importance of long-term emotional support from someone who had already survived the sudden loss of a loved one, she formed a non-profit peer-to-peer bereavement support organization, AirCraft Casualty Emotion Support Services (ACCESS), in the fall of 1996. ACCESS pair their grief mentors, who have all lost loved ones in past air disasters, with callers seeking grief support for more recent losses. What makes ACCESS unique is that they pair mentors with callers who have lost a similar relationship – for example, they pair mothers with mothers, fathers with fathers, orphans with orphans, siblings with siblings and so on with other relationships lost. ACCESS arrange for callers to talk with mentors who have been through a similar loss. ACCESS have received many testimonials about the significance of conversations with grief mentors. As word spread about the help they provided, ACCESS received more and more calls for help

Shortly after TWA 800, Swissair 111 and Egyptair 990 crashed out of JFK, followed by Alaska Airlines 261, the events of 9/11 and hundreds of other commercial, private and military air disasters. Resulting from these incidents, the need for ACCESS's services grew exponentially. ACCESS now has a team of 250 trained grief mentors on-call, who answer hundreds of calls for help each year. In addition, ACCESS focuses on preparations before an accident and trains first responders and personnel from airlines all over the world on how to best respond to those affected by air disasters. ACCESS has applied its model of grief support beyond aviation disasters to others types of sudden loss – including gun accidents and train derailments. The common factor is that the losses are unexpected – victims often leave for work and never return. 'We focus on rare things that are not supposed to happen,' says Snow.

All ACCESS grief mentors have the unenviable credential of having suffered the tragic loss of a loved one in an air disaster. ACCESS professionally train their grief mentors. ACCESS trainers emphasize confidentiality and prepare mentors for the wide range of emotions that surface from callers experiencing life-altering losses. Prospective grief mentors are trained to be active listeners and limit their interactions to emotional support.

MEMORIALS AND RITUALS

One of the many tasks a support group often focuses on is establishing a memorial that can be a permanent, visual reminder of the tragedy. A memorial is usually placed close to where the event happened, such as in the grounds of the World Trade Center in New York or at the crash site of an airliner. However, in some instances this is physically or practically impossible, as was the case after flight AF 447 crashed in the ocean on its way from Rio de Janeiro to Paris.

It may seem like an easy task, but agreeing on a memorial after a tragedy can be demanding and time consuming. In many instances the process takes several years, as was the case with the crash of American Airlines Flight 191 near Des Planes in May 1979. It took 32 years before the memorial was in place ... (*Chicago Sun Times*, 2011). Finding (and agreeing on) an artist is another challenge, and getting enough funding is also time consuming. Discussions about a proposed memorial often involve heated debates among members of the support group, local authorities and neighbors, and sometimes the proposal ends up in court. That was the case in Norway, when neighbors near the island of Utoya rejected the quite large memorial that was suggested.

The Memorial Garden in London commemorating those who died in the September 11, 2001 terrorist attacks is an example of a memorial created in cooperation with next-of-kin. According to Jelena Watkins, who represented one of the families, 'For many

FIGURE 10.2 A memorial can be a sophisticated piece of art or a simple plaque, like this one on a tree in Khao Lak, Thailand. *Source*: Knut-Erik Pedersen.

relatives, this is the only place bearing the name of their loved ones as most families were left with no remains to bury' (Watkins, 2012). The web page of the Grosvenor Square Garden explains that 'It is a living memorial and a testament to the love and courage of those who lost family, friends or colleagues' (Royal Parks, n.d.).

Communication regarding a memorial needs to be handled with eloquence and dignity. It is important to involve those directly affected, and to be prepared for varying views and opinions.

Another concrete reminder of a tragic event can be a ritual or a ceremony. Often arranged on an anniversary day of the incident, it can bring together survivors, victims' families, responders and the public in a communal ritual that represents a concrete way of remembering what happened and the lives lost. Such rituals take lots of planning, and again it is important to involve those most directly affected. The ceremony often involves speeches and songs, reading the names of the victims and a minute of silence. There are many examples of the importance of ceremonies, such as the anniversary of the tsunami being observed with music and speeches at a memorial in Oslo in 2014, or the tenth anniversary of 9/11 at Ground Zero, where Paul Simon performed 'The Sound of Silence.'

> "For me, the beauty of the Memorial Garden and other similar rituals is their ability to shelter our ongoing pain and transmute painful memories into remembrance.
>
> Jelena Watkins

SITE VISITS

Victims' families often want to see for themselves where their loved ones died. Such a visit can help them understand what happened and learn details about the order of events, and can be a way of expressing grief through rituals and ceremonies. If possible, next-of-kin often want to be at exactly the spot where death occurred, but depending on what killed them, this is not always possible.

FIGURE 10.3 After the tsunami, Norwegian next-of-kin made makeshift memorials at the hotels where their loved ones had died. *Source*: Kjell Brataas.

As described in Chapters 2 and 3, the Norwegian government has extensive experience in bringing family members 'back to where it happened.' The first collective visit to a site of a tragic event was in 1986, when 16 soldiers died in an avalanche in Vassdalen in the northern part of Norway. A few days later, next-of-kin were invited to visit the area and attend a memorial service, and three months afterwards, they were given the opportunity to get an aerial view of the mountain where the avalanche happened. A study 23 years later proved that the victims' families found the visits to be useful because they got an overview of what had happened, they felt they were close to the deceased and they were able to meet parents in a similar situation (Kristiensen & Kristiensen, 2011).

A more recent study that was conducted after the terror attack in Norway in 2011 examined bereaved parents' and siblings' experiences of visiting Utoya. The findings suggest that

> visiting the site of death can be recommended for bereaved families after disasters and terror. Although a visit to the site of death can be stressful, it is our conclusion that the benefits outweigh the burdens.

> Such visits can be particularly important for persons who struggle with complicated grief reactions such as avoidance of the reality of the death, and/or maladaptive grief-related ruminations. Adequate preparations are necessary before visits are conducted.
>
> (Kristensen, 2015)

Bringing survivors and next-of-kin to the site of a tragic event is challenging when it comes to logistics as well as communication. There are many wishes and personal preferences to consider, and at the same time, security, local authorities and available resources are determining factors.

For planning a site visit abroad, these questions should be considered:

- Who will be invited?
- How will costs be covered?
- What support staff are needed?
- How close to the accident site will participants be allowed?
- Is there a multi-cultural audience to consider?
- Which faiths are represented?
- Where can a ceremony be arranged?
- How will we handle media?
- How will we communicate to participants during the visit?

From my own experience with organizing site visits to Phuket in Thailand and Utoya in Norway, I have found that certain parts should always be included in a site visit or a ceremony:

- Being able to put down flowers and candles near the place where the person died
- Speeches – by victims, government officials, first responders and so on
- Songs, music or cultural performances

IDENTIFICATION AND REMAINS

For someone who has lost a family member or a friend, it is imperative that the loved one is identified so that a funeral can take place. However, this may be a difficult or impossible task, as the body might have been washed out to sea (as during the tsunami) or otherwise be impossible to locate. In a typical plane crash of a two-seater, authorities can expect to recover somewhere between 700 and 1200 body parts (Teahen, 2012), and finding and identifying them can take months or years.

From 2001 to 2005, almost 20,000 remains were collected at the former site of the World Trade Center. So far, 1,640 of the 2,753 victims have been identified, and the work continues to this day. Notification of remains is still ongoing, meaning that so many years after the terror attack on 9/11 2001, relatives are still receiving news about identification. In many instances, identification has taken many years, which means that some funerals had empty caskets or few body parts and that families have been notified several times about remains being found from the same person.

Identification in Difficult Surroundings

Ideally, police experts and coroners work on identifying remains in a sterile environment with modern equipment. That was not the case after the tsunami in Asia, where up to 1,000 bodies were stored on the ground outside the Bang Muang Temple. In the tropical climate there was no cooling available, and for the first days the identification team had no running water or electricity. Eight days after the disaster, dry ice became available for cooling down the bodies, and the next day large containers with cooling inside were available for storing. (Rognum, 2005)

DEATH NOTIFICATIONS

In a very few, special instances, communications experts might be asked to advise or take part in informing family members that their loved one has died. Such a conversation will probably be one of the most challenging tasks in a communicator's career, and it obviously requires the right people and preparation.

Before going to the door, the person(s) giving the death notification should check and verify facts regarding the person in question and be prepared to provide details about where the body is and the state of belongings. Wording is important, and direct and plain language – such as 'dead,' 'died' or 'killed' – should be used instead of 'softer' expressions such as 'gone to heaven' or 'has left us.' Do not be afraid to use the name of the deceased.

The Federal Bureau of Investigation and Penn State University have developed a comprehensive guide that includes a video course and a two-page document on death notification called 'We regret to inform you' (FBIPennState). It includes suggested delivery statements that might be useful to have on hand in an emotional and often stressful situation:

- I am so sorry.
- This is harder than most people think.
- Most people who have gone through this react similarly to you.
- People can experience many different feelings at the same time.
- This is one of the most difficult times in your life.

And a few statements to avoid:

- I know how you feel.
- Time heals all wounds.
- You need to be strong.
- You will get over this some day.
- You will find closure.
- He didn't know what hit him.
- You don't need to know that.
- Think of all your memories.

Updating the Next-of-Kin Database

All organizations and employers should have an updated and accessible list of their employees' next-of-kin. It should include the names, telephone numbers and addresses of the family members – or 'significant others' – the employee wants to be notified if something happens.

Many organizations struggle with keeping their next-of-kin list updated, and in some instances a form is only filled out when a person starts working in a company. However, life situations and family relationships change, so it is important – for the employee as well as his or her next-of-kin – that the list is always up to date. One way of doing this is to dedicate five minutes during an annual job review session to update the form.

It doesn't help to have updated forms if they are only accessible in a binder in the office. In the case of evacuation or fire, there needs to be a way that HR personnel can find next-of-kin information in a hurry, so additional storage options should be considered – including saving information on a USB or in a cloud service such as Dropbox or Google Drive.

The Roselawn 20th Anniversary

Support groups and memorial services often come about as a result of a person's commitment and dedication. One example is the 20th anniversary of the crash of American Eagle Flight 4184. Jennifer Stansberry Miller was instrumental in organizing the memorial; here is her description of the event:

> The 20th stemmed from a simple gesture from a former National Transportation Safety Board investigator who responded to our accident. He offered to speak to families at our anniversary, sparking the idea for a larger service for our twentieth year. We (myself, along with other families and interested community members) spent nearly two years planning this two day event. We created a community based project team which worked on: an informational public forum; designed, raised funds, and created a stone monument near the crash site; held a roadside memorial service to dedicate the new monument; hosted a family/fellowship dinner following the service; arranged

all logistics and comfort provisions (hotel, food, road closures, mental health support) and lastly, worked with American Airlines to obtain financial support of our monument and discounted tickets for families returning to Roselawn, Indiana. It was a meaningful experience to be able to bring our families, responders, airline employees, and local community together again to hear about the changes prompted by the crash, death of our loved ones, and the legacy created by the family advocacy work, and above all, honor our bonds and friendship which have developed over the years.

Typically, our anniversary is marked by private moments by our families either at their home or crash site/common burial site at Calumet Cemetery in Merrillville, Indiana; for those wanting to come together with others, a small roadside services has been held at the time of impact each year, with larger services being held at the 10 year, 15 year and 20th. Personally, the 20th signified a milestone for me; completing my personal chapter with Flight 4184 and beginning to write a new one. I will always remain connected to Roselawn, our families, stories, and history will always be with me. Our 20th commemoration was a special event, and I am grateful we were able to have that moment.

Jennifer Stansberry Miller lost her brother, Brad Stansberry, in the accident near Roselawn, Indiana on October 31, 1994. She was part of the aviation disaster victims' family association that advocated for the Aviation Disaster Family Assistance Act of 1996 and has provided education to federal, international and state agencies, air carriers, and emergency response personnel on the significance of family assistance. She is a clinical social worker working as a crisis consultant and was recently a subject matter expert for the Transportation Research Board Airport Cooperative Research Program Report 171, 'Establishing a Local Coordinated Family Assistance Program.'

BIBLIOGRAPHY

Chicago Sun Times. (2011, October 15). Retrieved from www.dpparks.org/parks-facilities/lake-park/flight-191-memorial

Dix, A. E. (2014). *Collective Conviction*. Liverpool: Liverpool University Press

FBIPennState. (n.d.). We regret to inform you … *Deathnotification*. Retrieved from http://deathnotification.psu.edu/assets/pdf/DeathNotification_PocketGuide_9-15.pdf

Global Counterterrorism Forum. (2012). *Madrid Memorandum*. Retrieved from www.thegctf.org/Portals/1/Documents/Framework%20Documents/A/GCTF-Madrid-Memorandum-ENG.pdf

IATA. (2016, December). *iata.org*. Retrieved from www.iata.org/publications/Documents/crisis-communications-guidelines.pdf

Kristensen, D. (2015). *Psykologitidsskriftet*. Retrieved from www.psykologtidsskriftet.no/?seks_id=443114&a=2

Kristiensen, P., &. Kristiensen, H.-P. (2011). *Bereavement and disasters: Research and clinical intervention*. New York: Routledge

Los Angeles County. (2014, January 31). *Lacoa.org*. Retrieved from http://lacoa.org/PDF/HazardsandThreats/Annexes/LACo_FAC_Plan_May2014_Web.pdf

NTSB. (n.d.). *icao.int*. Retrieved from www.icao.int/Meetings/HLSC2015/Documents/Presentations/Family%20Assistance-2-USA.pdf

Rognum, T. O. (2005, March 17). *Tidsskriftet*. Retrieved from http://tidsskriftet.no/2005/03/leder/identifisering-av-lik-etter-flodbolgen

Royal Parks. (n.d.). Retrieved from www.royalparks.org.uk/parks/grosvenor-squaregarden/things-to-see-and-do/september-11-memorial-garden

Teahen, P. R. (2012). *Mass fatalities: Managing the community response*. Boca Raton: CRC Press

Victim Support Europe. (2015, October 29). *Leaflet on Victim Support Europe*. Retrieved from http://victimsupport.eu/activeapp/wp-content/files_mf/1446120943WEBVSETrypticLeaflet210x99mm.pdf

Voices of September 11th. (2014). *Voicesofsept11.org*. Retrieved from http://voicesofsept11.org/initiatives/preparing-after-resource-kit

Watkins, J. (2012, November 21). *Mental Health Today*. Retrieved from www.mentalhealthtoday.co.uk/the_value_of_disaster_memorials_and_rituals_9675.aspx

CHAPTER 11

Preparing for the Worst

How an organization handles a crisis has a lot to do with planning. There will, of course, be many uncertainties and occurrences that no one can foresee, but having thought about what can go wrong – and how to handle the situation – *before* something happens is paramount to maintaining control over a situation. And with social media used by the public as well as government agencies and customers, crisis communication planning is more important than ever.

> "Good risk communication can rally support, calm a nervous public, provide needed information, encourage cooperative behaviors, and potentially help save lives.
>
> U.S. Environmental Protection Agency (Minamyer, 2007)

THE DISASTER COMMUNICATION CYCLE

Ideally, the steps for planning crisis communication should follow the spiral in the model shown in Figure 11.1. Starting with 'Business as usual,' the team would plan for scenarios and eventualities, then train and complete exercises, followed by an evaluation process. The result will be an improvement, which leads to greater readiness and back to business as usual, but at a 'higher' level of competence. However, if a crisis occurs, the organization moves from business as usual into crisis mode, when plans are put into practice. After the event, smart organizations evaluate their work in the form of systematic surveys or a large or small After Action Report. The feedback provides an opportunity to make improvements, and you achieve a higher level of readiness.

FIGURE 11.1 'The disaster communication cycle' explains how planning and training elevate the knowledge and readiness of an organization, so that it constantly evolves and improves.
Source: Kjell Brataas.

Planning should be done in concert, as a 'joint operation' with several departments within an organization. Top management, communications teams, HR and IT are key players, and they should meet frequently to go through the organization's preparedness and to evaluate – based on recent events and lessons learned – the existing plans and procedures. Planning for crisis communication should focus on practical questions, such as:

- Who speaks on the company's behalf when the CEO is unavailable?
- Who can change the top banner of the organization's Facebook page?
- How do we print something if we need to operate from an alternative location?
- Who can help us?
- How do we find out whether our employees are safe?
- How do we warn staff if they are in danger?
- Do we have protocols in place for alerting decision-makers?
- Where do we direct employees in the case of a physical threat, and who makes that decision?
- How do we handle a 'dynamic lockdown' situation?
- How do we communicate with employees during and after an evacuation?
- How do we coordinate with our most important stakeholders and partners?

According to crisis management expert Ken Jenkins, 'Nothing fails faster than a plan no one knew anything about.' How the plan is distributed and its 'location' therefore need to be thought of, and all members of the crisis management team should know the contents of the plan and where to find it in the case of an emergency. Paper binders are still useful, especially if digital means are unavailable or if you need to grab a document and run, but an updated digital version of the plan should also be available on company intranets, as a document in a closed Facebook group, or stored in a cloud service such as Dropbox or Google Drive.

SCENARIOS

A beneficial approach is to consider what can happen. To do this, you need to be a historian as well as a meteorologist: You must look at previous events in combination with a weatherman's foresight into what tomorrow will bring. Many scenarios will be specific to an organization's products or location, but there are several, more common, events that should be included in a list of what to prepare for:

- Fire
- Weather-related events (hurricanes, tornadoes etc.)
- Other natural events (flooding, earthquake etc.)
- Traffic accidents involving employee(s)
- Workplace violence
- Terror

There will be numerous communication challenges with each of these scenarios, and as mentioned earlier, the challenges will multiply if the crisis takes place in a foreign country or involves several nationalities.

Employees will react better, and lives can possibly be saved, if scenarios and their consequences are known to staff. Informing employees about possible threats and how they will be handled can be accomplished in a variety of ways, such as during an all-employee meeting, at department staff meetings, through internal news channels such as newsletters or intranets, or through digital platforms such as e-learning programs and closed Facebook groups. Emergency plans and crisis communication plans should be revised annually (or ideally more frequently), and there should be routines in place for making sure everyone who needs it has access to – and has read – the latest version.

Based on scenarios and internal discussions, several crisis communication tasks can be planned for and described in advance:

- Personnel and extra personnel
- Information flow from situation room to communications staff
- Premade black or grey headings for social media platforms
- Templates for invitations to press briefings

- Generic messages for Twitter
- Contact lists for important stakeholders
- A primary and a secondary spokesperson
- Tools for cooperation and sharing information (digital whiteboards, team discussion forums, conference calls, closed Facebook groups etc.)
- Social media accounts, alternate email systems (e.g. Google Mail, Hotmail etc.), and services for file sharing (Dropbox, Google Drive etc.)
- Contact list for journalists

Although mostly a positive part of handling a crisis, volunteer and donation management can become a challenge. Virtual volunteers (described in Chapter 6 on social media) can be a valuable asset, but more concrete forms of volunteering might require another kind of attention. There are many examples of communities or companies receiving hundreds of flowers, candles or teddy bears shortly after a tragedy, and all such items need to be handled

FIGURE 11.2 Even if 'cash is king,' when donations are needed after a disaster, many people will send clothes, teddy bears and so on. All these items need to be sorted and organized, as was the case in Texas after Hurricane Harvey. The picture shows the Aransas County Donation Center in Rockport in September 2017.
Source: Sarah K. Miller, Field Innovation Team.

properly. Communicating about such volunteerism can therefore become an extra task to handle, and one that can and should be prepared for in advance of a tragic event.

The donation achievements of the Boston Marathon bombings and the Pulse mass shooting are examples of excellent contribution management. Their success had a lot to do with an early start, and to be as effective as possible, a fund should be initiated within one to two days. Although its initiation is usually handled by others, informing people about such a fund and spreading the word definitely need the assistance of communications experts.

Choosing the Right Words

Not everything can be planned or agreed on in advance, but when a crisis situation occurs, it helps to have thought about what *words* to use when communicating with various stakeholders. To this end, the Australian Government in 2008 published a document called *Emergency warnings – choosing your words* (Australian Government, 2008). The guide describes specific words that according to research best convey the relevant concepts for warning messages. The document also points out that most crisis communication situations should be looked on as a dialogue, and not a command situation: 'Your role is not that of a "King" to issue orders – a better analogy is to be a marketer who is selling the product of "appropriate action" and needs to convince the audience to accept the advice.'

THE CRISIS COMMUNICATION PLAN

A crisis communication plan should be short and to the point, but it needs to include enough information so that people who do not work in the field of communications (such as colleagues in HR or IT) can read and understand it. A good start is a brief description of the organization's goals and principles for communicating in a crisis, followed by a list of scenarios that might necessitate the use of the document. Other parts of the plan might include:

- Notification and alerting protocols
- Quick response list
- Crisis communications team
- Key audiences
- Location for communication operations center
- Communication channels
- Crisis coverage review (traditional and social media), story trends and key stakeholder issues
- Contact lists (employees, media, police, government agencies, subject matter experts etc.)
- Technical issues (log-in procedures, wi-fi access etc.)

The first few hours after a crisis has occurred are usually hectic, and it is therefore of great help if the crisis communication plan includes an easy-to-read quick response list of what to do during the first few hours. Such a list could look like this:

- Notify CEO, colleagues and others who will be handling the situation.
- Call in extra staff.
- Start social media monitoring.
- Gather facts.
- Delete or delay scheduled posts.
- Inform the reception and/or switchboard.
- Develop a 'who answers what' with outside partners or agencies.
- Consider making preparations for a press conference.
- Develop FAQs.

Preparing a 'Grab Bag'

Several organizations, private corporations and government offices have preplanned who will travel in case of emergencies and what equipment they need to bring. A typical 'grab bag' can include items such as chargers, pens, USBs, headsets, an extra phone and a variety of batteries.

SAVING LIVES THROUGH BLEEDING CONTROL

Informing employees about what to do if a colleague or a bystander gets hurt is vital in saving lives. Too often, people near a bleeding victim will hesitate to do something, either because they want to wait for professional doctors or because they are afraid of making matters worse. The Hartford Consensus Group (a collaborative committee with members from law enforcement, trauma surgeons and emergency responders) has addressed this reluctance and coined the phrase 'If you see something – do something' (Bleedingcontrol.org, n.d.). Their goal is to improve the chance for survival of gunshot and mass casualty victims. 'Most shooting events are over in 15 minutes, and people can bleed to death within five minutes from severe injuries,' explains Dr Lenworth M. Jacobs, MD, Chairman of The Hartford Consensus. He points out that after a mass shooting or a terror attack, survival for many will pivot on the ability of trained civilians to tend to people's wounds while first responders are en route. There are several techniques for stopping bleeding. Ideally, a tourniquet kit should be used, but if such equipment is lacking, belts, shirts or other articles of clothing can be used effectively.

The Hartford Consensus addresses bleeding from any cause, not just gun shots or mass casualty events. Chainsaw injuries, knives or falling on sharp objects can cause severe bleeding, which should be stopped by immediate responders with the techniques advocated by the group.

TRAINING AND EXERCISES

There are many ways to train for the communication aspect of a crisis, ranging from discussions and 'table-top' exercises to full-scale drills with real actions. One type of exercise can focus on notification systems and alert-escalation, while another can test how various parts of an organization work together in a complex or destabilizing situation.

Conducting exercises has many merits:

- Preparing for an exercise often involves updating plans and procedures.

- An exercise is a 'safe' way of testing various parts of an organization in decision making in an unusual context.
- Taking part in an exercise creates a better understanding of the responsibilities and capabilities of various parts of an organization.
- An exercise strengthens personal relationships, which can prove valuable in the event of a real crisis.
- An exercise creates awareness that a crisis can occur.
- An exercise will reveal which resources are desired (and possibly need reinforcement).
- An exercise, especially if it includes social media, can test the handling of a deluge of information from a variety of sources.
- An exercise can highlight the need for adequate systems for virtual discussion boards, such as Slack, Trello, Facebook groups and so on.
- If the scenario is an internal event such as an active shooter or an accident within the office, an exercise can reveal who among the staff will be the first true responders.

Exercise Scenarios

For an exercise to work, the team behind it needs to come up with a scenario that participants find relevant and challenging. Typical scenarios deal with a fire in the building or a group of employees being involved in a traffic accident abroad. Other set-ups worth considering are a cyber-attack with massive loss of data, kidnapping of a senior manager or a social media scandal.

An exercise can last for only a couple of hours and take place in a regular meeting room with no special equipment or preparation and still be effective and produce actionable learning points. I call this the 'Germanwings approach.' It involves the CEO, the communications team and people from HR who meet for a couple of hours to discuss the following scenario: How would we react if four of our employees were on a business trip aboard a flight that recently crashed? Participants would then, as a table-top exercise, talk among themselves about responsibilities, messages, victim support, and internal and external communication. A few questions to start the discourse could be: 'Who calls family members of the employees?' 'What do we say internally?' 'Who should travel to the scene of the accident?' and 'What is the role of the CEO during the first hours of the crisis?'

There are several ways of making an exercise feel 'real.' Having 'fake' journalists bombard participants with challenging questions is a typical approach, possibly combined with an actor or actress pretending to be a distraught next-of-kin desperate for answers.

An easy and effective way of moving an exercise forward is to show participants made-up news stories about the crisis they are handling. Such stories can be presented as an update distributed on paper or a text message, or exercise staff can produce made-up news presented via PowerPoint.

Social media should always be a part of a crisis communication exercise, not only because they play such a prevalent role in today's handling of a crisis, but also because an exercise can focus on aligning internal and external messages. Having a 'flood' of questions and comments appear on social media during an exercise definitely increases realism and tests an organization's ability to answer appropriately. Social media surveillance, approval processes and virtual collaboration are all aspects of crisis communication that can be tested during an exercise that includes social media.

In a document titled 'Best Practices for Incorporating Social Media into Exercises,' the U.S. Department of Homeland Security describes reasons for incorporating social media in an exercise, including 'enhanced communications and engagement, enhanced situational awareness, and decision-making' (Department of Homeland Security, 2017). The report states that this kind of exercise 'allows an agency to replicate real-world actions in a safe, no-consequence environment to test responses to difficult situations.'

Simulation, Board Games and Role Plays

To achieve realism, several web services are available that can mimic social media messages and allow participants to take part in conversations on platforms similar to Facebook and Twitter. There are many advantages to this approach; perhaps the most important is that it enables realism and challenges on a closed platform that only participants can access. Using real Twitter or Facebook, or the company web page for crisis communication exercises, is never a good idea.

In addition to social media simulation, playing a board game can also function as an effective tool for training and exercises. Brooks Hogya, who has developed several games for disaster exercises, explains that

> Innovative approaches, like serious board games, can be used to bridge the gap between education and practice and give participants a space in which to explore complex academic ideas. The serious games, which offer players competition and cooperation, can also inform players about disaster risk reduction and community resilience.
>
> (Hogya, 2016)

Yet another approach to handling a crisis and making exercises more engaging is through think play. Calling herself a 'Chief Wrangler,' Desiree Matel-Anderson is the head of a non-profit called Field Innovation Team (FIT), which utilizes a three-step design process to decide how to expedite response and recovery after a crisis. Matel-Anderson explains:

> Whether the team is in the Syrian Refugee Crisis collaborating on an artificial intelligent chat-bot to provide psycho-social services in the Beqaa Valley of Lebanon, in the Northeast United States during Hurricane Sandy making decisive decisions to redesign survivor-centric recover centers, or curating a robotics petting zoo for Latin American youth in a border crisis, the team utilizes a three-step design process to decide how to expedite response and recovery utilizing games and design thinking.

The FIT team is also involved with climate change and disaster risk reduction, which includes working with global communities through games and design thinking.

Findings

Patrice Cloutier, a Canadian crisis communications expert, has suggested the following goals of crisis communication, which could be part of a crisis drill:

1. *Notify and alert*. Get the ball rolling immediately. Start occupying the public space.
2. *Put in motion your response plan*, but understand that scrutiny (from media and public) will be greater than ever before.
3. *Operationalize your social listening*. Get sound intel from social media to help decision makers understand public perception and how the crisis is evolving.
4. *Engage*. Keep communicating throughout the crisis after initial notification. And don't forget any internal audience.

A crisis communication exercise usually uncovers several points for improvement. They often shine light on borderline areas of responsibility, which can lead to productive discussions about who answers for what – internally as well as externally. Another common finding is the duration of meetings in the crisis management group – and how information about their decisions and discussions trickles down to the people responsible for taking action.

An evaluation of the exercise should also include discussions about the status of written plans and procedures, and whether they need updates or shortening. From a communication standpoint, other parts of an evaluation include the approval flow for issuing messages on Twitter and routines for designating a second-in-command who can function as a media spokesperson in case the CEO is not available or needs a break.

Situation reports are often challenging, especially if the crisis involves several entities providing intelligence. It can be quite challenging for an internal team to be on top of the situation, especially when reports from the police or government agencies contradict what is being reported by CNN or through social media.

Other findings can include:

- Not prioritizing victims
- Not setting aside enough financial resources

- The need for alternate locations and technical systems (if headquarters and email systems are unavailable)
- Change in top management during the crisis
- Unwillingness to have an honest evaluation

An exercise should also focus on the more practical side of handling a crisis, including adequate systems for logging and timekeeping, clean-up of meeting rooms and kitchens, and enough breaks, refreshments and sleep for participants.

> "
> The inconvenience and disruption created by full-scale drills is well recognized, but this can be ameliorated by having frequent, smaller, drills so that disruptions are kept at a minimum.
>
> 'Worksite emergency preparedness: Lessons from the World Trade Center Evacuation Study'
> (Burke & Cooper, 2008)

ACTIVE SHOOTER SCENARIOS

Active shooter scenarios or other mass casualty events are especially hard from a communication standpoint. One reason is that the situation is often extremely unclear, making it almost impossible to give concrete advice about where to go or how to avoid becoming a victim. Another factor is that such incidents usually are over before police arrive, putting an extra toll on crisis management and preparedness. According to trauma specialist Steven M. Crimando, the reality regarding mass shootings is that 'there are more events, involving great numbers of casualties, demanding a higher level of readiness' (Crimando, 2016).

The concept of one or more criminals or terrorists wanting to harm a large number of victims has many descriptions. 'Active shooter' is a term often used in the U.S., but 'Active threat' might be a better description. 'Workplace violence' often describes a disgruntled employee, while terrorist attacks in Nice and London show that 'vehicle ramming' is another threat to consider – and prepare for.

Run, Hide …

In addition to naming it differently, various countries and law enforcement organizations use diverse sets of advice for what to do if you get involved in an active shooter event. For a long time, law enforcement agencies in the U.S. have used three concrete verbs to instruct people what to do:

1. Run
2. Hide
3. Fight

In the U.K., the police have used a similar approach, but they have changed the last word from 'fight' to 'tell.' In an article in *Mailonline*, Detective Superintendent Scott Wilson explained what these three words mean in practice (Wilson, 2016):

> *Run* means don't hide under a table or play dead. Run out of the building. Don't try to negotiate.
>
> *Hide* means that if you can't get out of the building, barricade yourself behind a steel door, stay quiet and turn your phone on silent.
>
> *Tell* – phone the police as soon as you can, give exact locations, what kind of weapon the terrorist might have have had, are there any casualties, etc.

Although less confronting than its U.S. counterpart, the advice from the U.K. law enforcement does include an extra sentence explaining that 'if someone is in immediate danger and their life is being threatened we would never criticize their actions if instinct takes over and they feel the need to fight back' (City of London Police, 2015).

While three unpretentious words might seem easy to understand, communications experts and law enforcement personnel must realize that individuals caught up in an active shooter scenario might have problems following these simple rules for survival. Many victims will instead freeze, momentarily or for a longer period of time, which might make running or hiding impossible. Even worse, a short freeze might be followed by paralysis induced by panic. Lieutenant Colonel (Ret.) Mike Wood therefore suggests that the linear model should be changed:

> A proper model would make it clear that either running or hiding or fighting could be an appropriate initial response, depending on the circumstances. Teaching the public that violence can never be the first option fails to mentally prepare them for the times when it should be.
>
> (Wood, 2016)

Yet another approach comes from the Advanced Law Enforcement Rapid Response Training (ALERRT)™ Program at Texas State University. Their advice is to 'Avoid, Deny, Defend,' which includes having an exit plan, creating barriers or defending yourself by not fighting fairly (Avoiddenydefend.org, n.d.).

First Step: Hinder Active Shooter Events from Happening

The most effective advice for active shooter events is, of course, to make sure they don't happen. This is obviously easier said than done, but several safety measures should be taken at government and corporate offices. Peter Power, head of Visor Consultants, suggests the following precautions (Power, 2016):

- Restrict access to building areas.
- Immediately ground all lifts – without resorting to setting off fire alarms.
- Ensure you are able to quickly communicate with everyone in the building. This includes prerecorded (or at least prepared) messages.
- Reduce people becoming greater targets for an active shooter in the reception and/or ground floor where an attacker might use a previously seen tactic of breaking the fire alarm glass and thereby triggering an unnecessary evacuation. This means putting ground floor break glass on 'double-knock' where an

initial alarm is effectively silent for a brief while, allowing a double check and cancelling a false call if confirmed.

- Organize a rapid assessment and decision making.

If you can't avoid an active shooter event taking place, the next best step is to make sure that employees, students and the general public know what to do in case they are caught in the middle. Law enforcement agencies in many countries have therefore made various information materials about active shooter response available online and in print. The Department of Homeland Security funded a video about the 'Run, Hide, Fight' survival skills that received worldwide distribution, and several local agencies, corporations and police departments have produced posters and graphics that teach survival skills. In France, for example, large posters are on display at major tourist destinations advising individuals '*Vigilance attentat: les bons comportements*' – 'How to behave in case of a violent attack.'

'Invacuation' through Dynamic Lockdown

If victims of an active shooter event cannot escape out of the building, one alternative is to 'invacuate' – to hide in a locked part of the building. This kind of 'dynamic lockdown' might avert people from moving into a dangerous area where the shooter is, and at the same time prevent an attacker from getting further into a building. According to the U.K. National Counter Terrorism Security Office, 'the ability to frustrate and delay the attacker(s) during the course of the attack and reduce the number of potential casualties can be greatly increased through dynamic lockdown' (NaCTSO, 2015). Such a procedure involves many challenges, including how to inform the people in the building where to go and how to physically lock down the area. Important messages can be broadcast through a PA system, via text messages or as a predefined signal coming from alarm clocks in the building. The people who are hiding need to be told not to open any doors before police arrive and to constantly look for safe escape routes.

Planning for a dynamic lockdown obviously involves several departments of an organization, including HR, security and the building owners (who can assist with disabling elevators without returning them to the ground floor).

Instant, Global News

Even though it, sadly, happens regularly, an active shooter event becomes headline news as soon as it happens. This is not surprising, as it has all the ingredients of a thought-provoking story; a dramatic event, an unknown number of casualties, heroes and villains, questions regarding how it unfolded and whether it should or could have been avoided. Media attention spirals quickly, and staff as well as bystanders and witnesses need to know they do not have to speak to journalists. The shooting at the Sandy Hook High School in December 2012 is a telling example of what media interest can be like. According to Michael Cech, whose wife Yvonne was responsible for saving 21 people that day, media insensitivity became an issue because (Firestorm, 2015)

- Journalists began interviewing children without parental consent.
- The media were at every church and funeral following the shooting.
- Over 50 media outlets contacted Cech's family and arrived at his home within 48 hours.

It is not only the media that show up in large numbers after a mass shooting; government agencies will also play a role in its aftermath. Planning for handling such an event should therefore include a list of possible agencies that one will have to cooperate with when it comes to handling the crisis as well as communicating with media and the citizens. As an example, here is a list of the responding agencies after the Pulse shooting in Orlando in 2016:

- Orlando Police Department, Orange County Sheriff and multiple police agencies
- Orlando and Orange County Fire Departments
- Orlando Fire Dept. Arson/Bomb Squad
- Rural Metro (transport for Orlando)

- OPD SWAT
- Multiple hospitals
- Orange County Medical Examiner
- Florida Department of Law Enforcement
- Florida Highway Patrol
- Central Florida Intelligence Exchange
- Federal Bureau of Investigation
- Bureau of Alcohol, Tobacco, Firearms, and Explosives
- Drug Enforcement Administration
- Department of Homeland Security

OLTV in Norway

The Norwegian police have decided not to use any words connected with a weapon when describing a possible mass casualty event with a perpetrator. Instead, they use the description 'ongoing, life-threatening violence' – shortened to OLTV (or PLIVO in Norwegian).

BUSINESS CONTINUITY

Many companies and organizations have a plan for business continuity, which describes how they will function in the case of a major breakdown or crisis. Minimizing problems caused by inadequate communication and increasing the timeliness of messages are important factors in crisis management, and those who deal with business continuity therefore need to cooperate closely with those responsible for crisis communication. Working together, these professions can make sure critical communication solutions are in place so that crisis managers can be rapidly aware of an incident and how it unfolds.

Social media provide tools that are important also from a business continuity standpoint. A closed Facebook group can function as an internal discussion forum, while a Facebook page can be the company's main information hub in case the normal website crashes.

BIBLIOGRAPHY

Australian Government. (2008). *Emergency warnings – choosing your words*. Attorney-General's Department

Avoiddenydefend.org. (n.d.). Retrieved from www.avoiddenydefend.org/add.html

Bleedingcontrol.org. (n.d.). Retrieved from www.bleedingcontrol.org

Burke, R. J. and Cooper, C. L. (2008). *International terrorism and threats to security*. Cheltenham, U.K. and Northampton, MA, U.S.: Edward Elgar Publishing. Full report available at www.researchgate.net/publication/286706062_Worksite_emergency_preparedness_Lessons_from_the_world_trade_center_evacuation_study

City of London Police (2015). Retrieved from www.cityoflondon.police.uk/news-and-appeals/Pages/Public-given-advice-on-what-to-do-in-the-event-of-a-terrorist-attack-.aspx

Crimando, S. (2016). Retrieved from www.disaster-resource.com/index.php?option=com_content&view=article&id=2849&Itemid=166

Department of Homeland Security. (2017, March). *Homeland Security*. Retrieved from www.dhs.gov/sites/default/files/publications/Best-Practices-Incorporating-Social-Media-Into-Exercises-508%20.pdf

Firestorm (2015). Retrieved from www.firestorm.com/resource/form-download-brief-sandy-hook-follow-up

Hogya, B. (2016). Disaster risk reduction through play: A serious game based on research, comprised of research, for the purposes of research. Retrieved from http://resiliencebydesign.com/portfolio_page/resilienceville

Minamyer, S. K. (2007). *Effective risk and crisis communication during water security emergencies: Report of EPA sponsored message mapping workshops*. U.S. Environmental Protection Agency. Washington, DC: U.S. Environmental Protection Agency

NaCTSO. (2015). *gov.uk*. Retrieved from www.gov.uk/government/uploads/system/uploads/attachment_data/file/478003/NaCTSO_Guidance_Note_1_-_2015_-_Dynamic_Lockdown_v1_0.pdf

Power, P. (2016). *thebci.org*. Retrieved from www.bcifiles.com/Q12016_online.pdf

Wilson, S. (2016). Retrieved from www.dailymail.co.uk/news/article-3798690/They-trying-kill-know-survive-terrorist-attack-anti-terror-cop-reveals-tactics-adopt-escape-hiding-desk-really-bad-idea.html

CHAPTER 12

Psychological Reactions

There are hundreds of books that explain psychological behavior in times of crisis. This chapter will therefore touch on a few general topics that crisis managers and communicators should be aware of and that hopefully will make it easier to understand why people respond in sometimes unexpected ways in the midst of a tragedy.

NATURAL REACTIONS

Survivors of a tragedy – and family members of those who died – display a variety of psychological emotions, from crying and anger to no noticeable responses at all. These can take a long time to surface, and it is not uncommon for next-of-kin to have a first major reaction on the fifth or tenth anniversary of the tragedy. It is also a fact that large-scale traumas can take a lifetime to heal, and that certain events will make the tragedy resurface. The first anniversary, birthdays or Christmas celebrations can bring back memories of loved ones, and Mother's or Father's Day easily becomes a reminder of what could have been. Certain words can also prompt a psychological reaction, for example when survivors of a mass shooting hear everyday conversations about 'bullet points' or 'triggers.'

In addition to 'expected' behavior like crying and anger, victims sometimes have other reactions or fall-out from what they have experienced. Relationship issues are not uncommon, and the incidence of substance abuse and suicide also escalates for victims.

A study in Norway in 2012 showed that half the survivors from the island of Utoya (77 people died in the attack on Oslo and the island) had depressive reactions or post-traumatic stress. As one of them said in an interview, 'On Monday I will be admitted to a psychiatric hospital, so that I will not become number 78' (Fuglehaug, 2013).

Fortunately, there are several 'best practices' for taking care of victims or bereaved family members. In Norway, several studies have been conducted on the reactions and follow-up after the Utoya mass shooting. A study by Dr Atle Dyregrov and his team at the Center for Crisis Psychology in Bergen describes how proactive follow-up consisting of several weekend gatherings for bereaved family members proved helpful (Dyregrov, 2016). They gathered over four weekends to process and learn about grief, including group sessions, plenary lectures, workshops and social activities. The same structure was used in each gathering.

According to Dyregrov, the weekend gatherings had several objectives:

- To signal society's (and public authority's) care and concern for those involved
- To help many at the same time
- To mobilize mutual support and help within and between families
- To normalize and validate people's experiences and reactions (a psychoeducative component)
- To provide access to important information related to the event and its consequences
- To help integrate the loss into their future lives by supporting family members' own coping mechanisms, and inform/teach them self-help methods
- To secure early identification and referral of those in need of individual assistance

The four weekends had different themes, including subjects such as grief over time, the court case and family interaction. Resilience was a major emphasis:

> In addition to leaving ample time to talk about grief, sorrow and longing, the focus through all four gatherings was very much on how

> they could develop and learn coping methods to have the best foundation to handle their daily function.
>
> (Dyregrov, 2016)

Right after a tragedy, people and communities often come together, and there is an enormous sense of care. We have seen this after 9/11 and after the Manchester bombing, when strangers helped each other and neighbors spoke for the first time. However, this sense of togetherness seldom lasts. Sooner or later reality sets in, as individuals and groups compete for resources and next-of-kin and survivors fight for attention. The latter group might have their own psychological challenges after having been through a traumatic event, but they are sometimes told they do not have the right to react or get help, as 'At least your person is still alive.'

THE PSYCHOLOGY OF EVACUATIONS

Psychological reactions to life-threatening situations vary greatly. Some (but usually only a few) people panic, while others take the lead and provide assistance. Unfortunately, many fail to do enough to save their own lives. According to Dr John Leach, a psychologist in the Extreme Environmental Medicine & Science Group at the University of Portsmouth, human behavior in a stressful environment can be summed up with the numbers 75, 15 and 10 (Leach, 1994). According to Dr Leach's research, 75% of people will carry on as though nothing has happened or will do nothing and wait for orders, 15% will do the right thing immediately and 10% will act in a way that is contrary to survival. Giving clear and direct orders is therefore of utmost importance in a crisis situation, as lives can be saved by immediately instructing possible victims in what to do.

Quite often, though, the first stages of a tragedy are chaotic and bewildering, and people in harm's way sometimes believe that the 'bangs' they are hearing come from fireworks or are part of a show, when it's actually *bullets* that are making the sounds.

A study from 2008 that looked at the evacuation of the World Trade Center on 9/11, 2001 found that a few employees in the buildings delayed their evacuation (Burke & Cooper, 2008). There were several contributing factors, for example lack of information about what was going on and the severity of the situation. According to the study, 'another factor that led to delays in initiation was the performance of tasks.' Employees took time to make phone calls, collect personal items, save computer files or complete last-minute work activities before evacuating. When they did start the climb down the many flights of stairs to safety, some people were also hindered by their footwear, as high heels and slip-ons made descent slow and difficult. However, the report states: 'By far the most important factor, at the organizational level, that affected both the initiation and the progression of evacuation, was leadership.' Lack of trained leadership led to delays, and in some instances staff did not know the location of emergency exits.

NEXT-OF-KIN PRIORITIES

Lessons learned from a number of tragedies show that some aspects of a crisis are more important than others. Naturally, *identifying the dead* is at the top of the list, but unfortunately this cannot always be accomplished. Next-of-kin also value *personal belongings*, and those handling a crisis should therefore be aware of how important it is to return an item that belonged to the deceased. A third factor is *recognition*, as it is more important than people from the 'outside' realize for victims to be acknowledged.

In her book *Surviving sudden loss*, Heidi Snow presents common threads from victims from a number of tragedies (Snow, 2012). They are worth bearing in mind from a communications as well as a human relations standpoint:

- Vivid memory of their last interaction with loved ones
- Notification: Suddenly life stops
- Hope for reappearance
- Open questions
- Immediate plans change
- 'The vanishing future'
- Regrets
- Loneliness and isolation
- Pain
- Self-doubt – grief can last a long time
- A life-long journey

- Changed relationships and new friends
- Survival is possible

'AREN'T THEY OVER IT YET?'

Many victims experience a social norm expecting them to get on with their lives sooner rather than later after a tragedy has occurred. However, most victims find that closure is a concept that is unobtainable following a death, and that neither a funeral nor a first anniversary will end the suffering or the memories of someone they loved. According to Michelle Steinke-Baumgard, founder of One Fit Widow, the West has the concept of grieving all wrong. In an article in *Huffington Post*, she explains that 'Western society has created a neat little "grief box" where we place the grieving and wait for them to emerge fixed and whole again' (Steinke, 2016). Steinke-Baumgard describes the grief box as small and compact, and continues to offer these truths behind life after loss (reprinted with permission):

Expectation: Grief looks a certain way in the early days. Tears, intense sadness, and hopelessness.

Reality: Grief looks different for every single person. Some people cry intensely, and some don't cry at all. Some people break down, and others stand firm. There is no way to label what raw grief looks like as we all handle our loss in different ways due to different circumstances and various life backgrounds that shape who we are.

Expectation: The grieving need about a year to heal.

Reality: Sometimes grief does not even get started till after the first year. I've heard countless grieving people say year two is harder than year one. There is the shock, end of life arrangements and other business matters that often consume the first year and the grieving do not have the time actually to sit back and take the time to grieve. The reality is there is no acceptable time frame associated with grief.

Expectation: The grieving will need you most the first few weeks.

Reality: The grieving are flooded with offers of help the first few weeks. In many cases, helping the grieving six months or a year down the line can be far more helpful because everyone has returned to their lives and the grief stricken are left to figure it out alone.

Expectation: The grieving should bury the dead forever. After a year, it is uncomfortable for the grieving to speak of their lost loved one. If they continue to talk about them, they are stuck in their grief and need to 'move on.'

Reality: The grieving should speak of the dead forever if that's what they wish to do. When someone dies, that does not erase the memories you made, the love you shared and their place in your heart. It is not only okay to speak of the dead after they are gone, but it's also a healthy and peaceful way to move forward.

Expectation: For the widowed – If you remarry you shouldn't speak of your lost loved one otherwise you take away from your new spouse.

Reality: You never stop loving what came before, and that does not in any way lessen the love you have for what comes after. When you lose a friend – you don't stop having friends, and you love them all uniquely. If you lose a child and have another, the next child does not replace or diminish the love you had for the first. If you lose a spouse, you are capable of loving what was and loving what is … one does not cancel out or minimize the next. Love expands the heart, and it's okay to honor the past and embrace the future.

Expectation: Time heals all wounds.

Reality: Time softens the impact of the pain, but you are never completely healed. Rather than setting up false expectations of healing let's talk about realistic expectations of growth and forward movement. Grief changes who you are at the deepest levels and while you may not forever be in an active mode of grief you will forever be shaped by the loss you have endured.

Expectation: If you reflect on loss beyond a year you are 'stuck.'

Reality: Not a day goes by where I am not personally affected by my loss. Seeing my children play sports, looking at my son who is the carbon copy of his Dad or hearing a song on the radio or smell in the air. Loss becomes part of who you are and even though I don't choose to dwell on grief it has a way of sneaking in now and again even when I'm most in love with life at the current moment. It's not because we dwell or focus, and it's not because we don't make daily choices to move forward. It's because we loved and we lost, and it touches us for the remainder of our days in the most profound ways.

Expectation: When you speak of the dead you make the griever sad, so it's best not to bring them up.

Reality: When we talk about our lost loved one we are often happy and filled with joy. My loss was six and a half years ago and to this day, my late husband is one of my favorite people to talk and hear about. Hearing his name makes me smile and floods my mind with happy memories of a life well lived. It makes the grieving sadder when everyone around them refuses to say their name. Forgetting they existed is cruel and a perfect example of our stifled need to fix the unfixable.

Expectation: If you move forward you never loved them or conversely if you don't move forward you never loved them.

Reality: The grieving need to do what is right for them, and nobody knows what that is except the person going through it.

Expectation: It's time to 'move on.'

Reality: There is no moving on – there is only moving forward. From the time death touches our lives we move forward, in fact, we are not given a choice but to move forward. However, we never get to a place where the words move on resonate. The words 'move on' have a negative connotation to the grieving. They suggest a closure that is nonexistent and a fictitious door we pass through.

Expectation: Grief is a linear process and a series of steps to be taken. Each level is neatly defined and the order predetermined.

Reality: Grief is an ugly mess full of pitfalls, missteps, sinking, and swimming. Like a game of shoots and ladders, you never know when the board might pull you back and send you down the ladder screaming at the top of your lungs. Just when you think you've arrived at the finish, you draw a card that sends you back to start and just when it appears you've lost the game you jump ahead and come one step closer to the front of the line.

Expectation: The grieving should seek professional forms of counseling exclusively.

Reality: The grieving should seek professional forms of counseling but also the grieving should look strongly towards alternative modes of therapy like fitness, art, music, meditation, journaling and animal therapy. The grieving should take an 'active' part in their grief process and understand that coping comes in many different forms for all the different people who walk this earth.

Expectation: The grieving either live in the past or the present. It is not possible to have a multitude of emotions.

Reality: The grieving live their lives with intense moments of duality. Moments of incredible happiness mixed with feelings of deep sadness. There is a depth of emotion that forever accompanies those who have lived with a loss. That duality can cause constant reflection, and a deeper appreciation of all life has to offer.

Expectation: The grieving should be able to handle business as usual within a few weeks.

Reality: The brain of a grieving person can be in a thick fog, especially for those who have experienced extreme shock, for more than a year. Expect forgetfulness, a reduced ability to handle stress and grayness to be commonplace after a loss.

BIBLIOGRAPHY

Burke, R. J. and Cooper, C. L. (2008). *International terrorism and threats to security*. Cheltenham, U.K. and Northampton, MA, U.S.: Edward Elgar Publishing. Full report available at www.researchgate.net/publication/286706062_Worksite_emergency_preparedness_Lessons_from_the_world_trade_center_evacuation_study

Dyregrov (2016). Weekend gatherings for bereaved family members after the terror killings in Norway in 2011. *Bereavement Care*, *35*(1), 22–30

Fuglehaug, W. (2013, January 7). *Aftenposten*. Retrieved from www.aftenposten.no/norge/Jeg-er-sliten_-og-trenger-hjelp_-men-jeg-skal-ha-mitt-liv-tilbake-132164b.html

Leach, J. (1994). *Survival psychology*. Basingstoke: Palgrave Macmillan

Snow, H. (2012). *Surviving sudden loss*. Self-published

Steinke, M. E. (2016, March 6). *Stifled grief: How the West has it wrong*. Retrieved from www.huffingtonpost.com/michelle-e-steinke/stifled-grief-how-the-wes_b_10243026.html

CHAPTER 13

Additional Information and Further Reading

Below is suggested reading on subjects covered in this book:

The unthinkable: who survives when disaster strikes – and why by Amanda Ripley (Three Rivers Press).
Resilience by Ken Jenkins (SEGR Publishing).
Surviving sudden loss by Heidi Snow (self-published by the author).
On crisis communication by James E. Lukaszewski (Rothstein Associates Inc.).
Collective conviction – the story of disaster action by Anne Eyre and Pam Dix (Liverpool University Press).
The crash detectives by Christine Negroni (Penguin Books).
Disaster heroes by Suzanne Bernier (Faith Books & MORE).

Index